Yung-Chen Lu

Singularity Theory
and an Introduction to
Catastrophe Theory

D1379480

Springer-Verlag
New York Heidelberg Berlin

Dr. Y. C. Lu

Ohio State University

Department of Mathematics
The Ohio State University
Columbus, Ohio 43210

AMS Subject Classification: 55G37, 55G99, 57D45, 55D99, 58Exx, 58F99

ISBN 0-387-90221-X Springer-Verlag New York Heidelberg Berlin
ISBN 3-540-90221-X Springer-Verlag Berlin Heidelberg New York

Library of Congress Cataloging in Publication Data:
Lu, Yung-Chen, 1938–
Singularity Theory and an Introduction to Catastrophe Theory
(Universitext)
Bibliography: p. 192
Includes index.
1. Differential Mappings. 2. Singularities (Mathematics).
3. Catastrophes (Mathematics). I. Title.
QA613.64.L8 514'.7 76-48307

All rights reserved.

No part of this book may be translated or reproduced in any form without written permission from Springer-Verlag.

© 1976 by Springer-Verlag New York, Inc.

Printed in the United States of America

PREFACE

In April, 1975, I organised a conference at the Battelle Research Center, Seattle, Washington on the theme "Structural stability, catastrophe theory and their applications in the sciences". To this conference were invited a number of mathematicians concerned with the mathematical theories of structural stability and catastrophe theory, and other mathematicians whose principal interest lay in applications to various sciences - physical, biological, medical and social. René Thom and Christopher Zeeman figured in the list of distinguished participants.

The conference aroused considerable interest, and many mathematicians who were not specialists in the fields covered by the conference expressed their desire to attend the conference sessions; in addition, scientists from the Battelle laboratories came to Seattle to learn of developments in these areas and to consider possible applications to their own work. In view of the attendance of these mathematicians and scientists, and in order to enable the expositions of the experts to be intelligible to this wider audience, I invited Professor Yung-Chen Lu, of Ohio State University, to come to Battelle Seattle in advance of the actual conference to deliver a series of informal lecture-seminars, explaining the background of the mathematical theory and indicating some of the actual and possible applications. In the event, Yung-Chen Lu delivered his lectures in the week preceding and the week following the actual conference, so that the first half of his course was preparatory and the second half explanatory and evaluative.

These lecture notes constitute an expanded version of the course. They are quite self-contained except with regard to certain proofs;

in particular, the material may be read and understood by those who are not familiar with the conference proceedings.* On the other hand, they should by very valuable to those non-specialists wishing better to understand the theory and applications of structural stability and catastrophe theory, as treated for example by the contributors to the conference proceedings. Although several very good expositions of the mathematical theory have recently appeared, I believe that these notes of Yung-Chen Lu are unique with respect to their very elementary starting point in mathematical analysis, and their emphasis on motivation and applications. I therefore strongly commend them, and wish to take this opportunity to thank my friend Yung-Chen for his very special contribution, through his lectures, to the success of the Battelle conference.

Peter Hilton

Battelle Seattle Research Center.
September 1976

*Published as Lecture Notes in Mathematics, Vol. 525, Springer-Verlag (1976)

CONTENTS

CONTENTS

INTRODUCTION

These lecture notes are based on the six lectures the author presented at Battelle Seattle Research Center in April, 1975. The lectures were given in the context of the Symposium on Structural Stability, Catastrophe Theory and Their Application in the Sciences (April 21-25, 1975), which also took place at the Battelle Seattle Research Center, and were intended to provide background material necessary for a deeper understanding of the more specialized presentations at the Symposium [68].

The general audience for these lectures thus did not consist of experts in singularity theory. The author therefore wishes to emphasize that these notes are intended for beginners in this field of mathematics and for those scientists who wish to get some flavor of this newly-developed mathematical theory. It is for this reason that the presentation of the material in these notes has been arranged in so elementary and comprehensive a way that even strong undergraduate students should be able to understand most of the contents. There are plenty of examples and, most importantly, intuitive and geometrical descriptions of fundamental concepts have been presented before the complicated formal definitions of such concepts; for example, universal unfolding, k-transversality, etc.

Zeeman wrote in the beginning of his article [100] "Catastrophe Theory" that: "Things that change suddenly, by fits and starts, have long resisted mathematical analysis. A method derived from topology describes these phenomena as examples of seven 'elementary catastrophes'." There are two quite distinct forms which 'catastrophe theory' assumes. One is the mathematical interpretation; the other consists of applications either to different areas of mathematics or to the mathematical analysis of a physical problem. As a mathematical discipline, catastrophe theory in its usual form is merely a classification theorem of degenerate singularities of codimension less than or equal to four, the techniques of which use unfolding theory extensively, as well as a crucial observation made by Thom [74], commonly called the Splitting Lemma or Theorem of the Residual Singularity. The second form, wherein applications are studied has resulted in various developments, including (a) some very interesting contributions to caustics of solutions of linear partial differential equations by Guckenheimer [22], (b) caustics of propagation phenomena by Jänich [25] and (c) some very interesting contributions to biological science as well as sociological science by Zeeman [95, 96, 100]. There are also two mathematical offshoots by Wasserman [83] and Baas [5] which are attempts at expanding or generalizing the "classical" or original formulation by Thom. But an appreciation of them requires first an understanding of Thom's original set-up.

Thom's book "Stabilité Structurelle et Morphogénèse" is hard to read for both mathematicians and biologists. The basic theme is quite as much philosophical as it is mathematical or scientific, since it is concerned to treat the very nature of the process whereby mathematics may be used to provide a model of the changing nature of the real world. As John Guckenheimer has written in his very perceptive review of the book [23], "René Thom has written a provocative book [74]. It contains much of interest to mathematicians and has already had a significant impact upon mathematics, but [it] is not a work of mathematics." Thus there remains the practical problem of how mathematicians who are expert neither in the principal areas of mathematics utilized by Thom nor in the currently dominant fields of application can acquire a knowledge of the subject adequate for a genuine comprehension. This practical problem is the motivation for these lecture notes.

A few words about the contents of these lecture notes may be in order. First, it must be said that these notes by no means provide a full exposure to singularity theory, but rather provide a means of entering the theory.

The first chapter is based on an introductory lecture and contains some elementary definitions, examples and some historical remarks. Some of the examples in this chapter should be studied again after Chapter 3.

The second chapter contains a rather detailed study of Hassler Whitney's landmark paper "On Singularities of Mappings of Euclidean Spaces I. Mappings of the Plane into the Plane" [87]. The author feels that this paper should be the starting point for whoever would like to study singularity theory. Basic definitions such as jet space, transversality, etc. are given at the beginning of the chapter. They are the essential ingredients for proving the genericity part of the Morse Lemma and Whitney's Theorem in the paper mentioned above. The author has no intention of covering all the details in H. Whitney's paper, but rather has sought to provide examples, geometrical explanations and a certain amount of mathematical derivation to lead the reader to comprehend the basic ideas of the paper.

The third chapter provides a study of finite determinacy and of the universal unfolding of singularities. The author refers the reader to [10, 82, 98] for further study in these subjects. However, the examples and the intuitive descriptions relating to the definitions involved in this chapter should help the reader to understand the material presented here. We have provided a proof of a special case of the Malgrange Preparation Theorem. The reason for doing so is to enable the reader to comprehend the spirit of the proof of this theorem without getting bogged down in details. The second part of this chapter is intended to establish the relationship between this theorem and the study of universal unfoldings. The main result is to provide the standard form of a universal unfolding of a singular map-germ.

The fourth chapter is an introduction to catastrophe theory. The author uses many examples to illustrate how the mathematical model is appropriate to various natural phenomena. As Peter Hilton has written in the introduction to his article "Unfolding of Singularities" [24], "to prepare a student to work in Thom's theory, it is clear that one must train him (and oneself) to become a genuine mathematician, but one imbued with the desire to use mathematics to understand the nature of the real world." This represents the author's intention in writing this chapter.

The fifth chapter describes H. Whitney's stratification theory. Basically this is a topic in algebraic geometry, however it is of fundamental utility in dealing with the problems of singularity theory.

Finally, the last chapter is an exposition of C^0-sufficiency of jets, another mathematical concept which is fundamental in discussing the problems of singularity theory.

There are two appendices, the first one looks again at Thom's Classification Theorem and explains Thom's three basic principles in Morphogenesis. The second one gives a proof of Thom's Classification Theorem.

ACKNOWLEDGEMENT

I am grateful to Battelle Seattle Research Center and many of its staff members for making it possible to arrange this lecture series. Many thanks are also due to T. C. Kuo and Benjamin Lichtin, who made a lot of valuable comments and suggestions for many of the formulations in the text, and to Professor René Thom for many useful conversations and for intellectual stimulation. My appreciation is extended to the Mathematics Department of The Ohio State University for technical assistance, to Miss Dodie Huffman for flawless typing, and to Mr. J.N.O. Moore for skillful drawings.

I am grateful to W. Kaufmann-Bühler and Larry Sirovich for their extensive assistance and courtesy during the preparation of this manuscript. I would like to thank the production and editorial staff of Springer-Verlag for their help.

Finally I am especially grateful to Professor Peter Hilton for the invitation to give this sequence of lectures at BSRC, for his careful reading of this manuscript from cover to cover, for his many valuable suggestions and, most importantly, for his invaluable encouragement and advice; in short, for having made this work possible. Thanks Peter.

Yung-Chen Lu
Columbus, Ohio
July 1976

CHAPTER 1

INTRODUCTION TO SINGULARITY THEORY WITH HISTORICAL REMARKS

1. Introduction with Naive Discussions

There are three basic ideas, which are related to each other in singularity theory. They are:

(1) Stability

(2) Genericity

(3) Unfolding of singularities.

For (1), let us consider the following two examples:

Example 1.1. Let $f: \mathbb{R} \to \mathbb{R}$ be given by

$$f(x) = x^2 \tag{1.1}$$

The graph of this map is as follows.

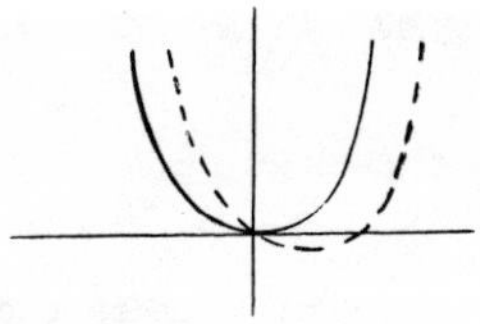

Figure 1.1.

Naively speaking this map is stable since if we push (or perturb) the graph of the map slightly (as shown by dotted line), the topological pictures of the dotted graph and the solid graph are the same. More rigorously, the dotted curve is just the graph of a reparametrization of f.

Example 1.2. (Whitney's cusp) Let $f: \mathbb{R}^2 \to \mathbb{R}^2$ be given by

$$f(x,y) = (u,v)$$

where

$$u(x,y) = xy - x^3$$
$$v(x,y) = y\,. \tag{1.2}$$

Whitney proved in [87] that f is stable in the following sense: if f is perturbed slightly in the C^∞-topology (this will be defined in section 2 of this chapter), the new map is just a reparametrization of f , or the new map has the same C^∞-type as f .

What we wish to do is to characterize differentiable maps that are stable. They are nice in the sense that when we pertrub them a little we can still predict their topological type.

The natural question after (1) is whether there are enough stable maps. In other words, can any map f_0 be approximated by a stable map? This is a typical question about genericity.

As to (3), the basic concern here is about unstable mappings. Unfolding is an important notion in singularity theory, introduced by René Thom. Let us investigate the following example:

Example 1.3. Consider the mapping $f: \mathbb{R} \to \mathbb{R}$ given by

$$f(x) = x^3\,. \tag{1.3}$$

This map is unstable at 0 (this statement will be proved in section 4 of this chapter). Intuitively we can see this fact in the following way. Perturb x^3 by ux, where u is small, the perturbed map $x^3 + ux$ assumes different topological types for $u < 0$ and $u > 0$, since for $u < 0$ we have two critical points in a small neighborhood of $(0,0) \in \mathbb{R}^2$ and for $u > 0$, we have none (see figure 1.2). Nevertheless the map $F(x,u) = x^3 + ux$ is "stable" (this is Whitney's cusp, to be discussed in Chapter 2) and f is "imbedded" in F (in

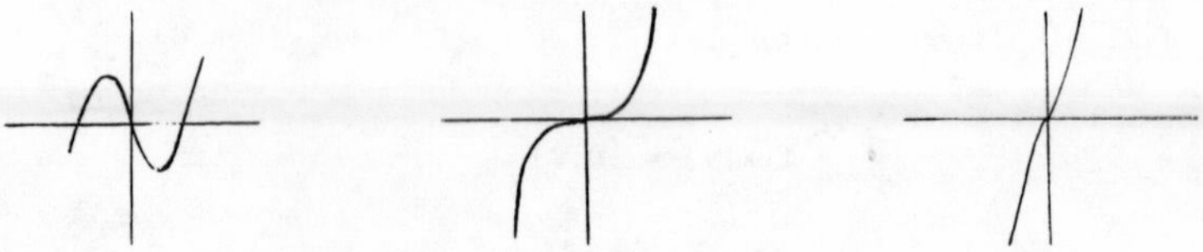

$u < 0$ $u = 0$ $u > 0$

Figure 1.2.

the sense $F|_{\mathbb{R}\times\{0\}} = f$). We say F is an unfolding of f . Thus the third question is: Given an unstable map f, how can one unfold f into a "stable" map (stable in the sense of unfoldings)? Moreover, we wish to do so in the most economical way, i.e. with the least number of parameters (like u in this example). This is the idea behind universal unfolding.

Morse and Whitney [54,55,87] were, of course, the initial developers of the subject of singularity theory. It is their work that one perceives the emphasis placed upon these three ideas. However, as we have already mentioned, the full utility of the third notion is a relatively more recent motif due to René Thom.

2. Elementary Definitions

For those who have the basic knowledge in topology and differential geometry could skip this section.

Let S be a set with a topology and an equivalence relation e . An element $x \in S$ is stable (relative to e) if the e-equivalence class of x contains a neighborhood of x . A property P of elements of S is generic if the set of all x in S satisfying P contains a set A which is a countable intersection of open dense sets. The following examples, from [81], can illustrate these two important concepts well. Let S be the vector space of complex $n \times n$ matrices. Let $G = GL(\mathbb{C},n)$ be the group of invertible complex $n \times n$ matrices.

Examples 2.1. Two elements x, y in S are equivalent if and only if there is an element $g \in G$ such that $y = x \cdot g$ under matrix multiplication. Thus x is stable if and only if x is invertible. In this example stability is a generic property.

Example 2.2. Two elements $x, y \in S$ are equivalent if there is an element $g \in G$ such that $y = g^{-1}xg$. The Jordan normal form of x represents its equivalence class in this case. A small perturbation in x will, in general, change the eigenvalue, so that no $x \in S$ is stable.

The above mentioned set A will be called the <u>residual set</u>. Examples will be found in section 3 and its successors.

<u>A differentiable n-manifold M</u> (or a n-dimensional differentiable manifold) is a locally euclidean space with differentiable patches. More precisely, in the first place, M is a topological manifold in the sense that it is a Hausdorff, second countable topological space and for each x in M, there is a neighborhood U of x in M together with a homeomorphism $\varphi\colon U \to \mathbb{R}^n$ onto an open set $\varphi(U) \subset \mathbb{R}^n$. The pair (U,φ) is called a <u>chart</u> at x. Now, M is a <u>differentiable manifold</u> if there is a collection of charts $\{U_\alpha, \varphi_\alpha\}$, called <u>atlas</u>, with $\cup U_\alpha = M$ and for any two charts in the atlas the map $\varphi_\beta \circ \varphi_\alpha^{-1}\colon \varphi_\alpha(U_\alpha \cap U_\beta) \to \varphi_\beta(U_\alpha \cap U_\beta)$ is differentiable.

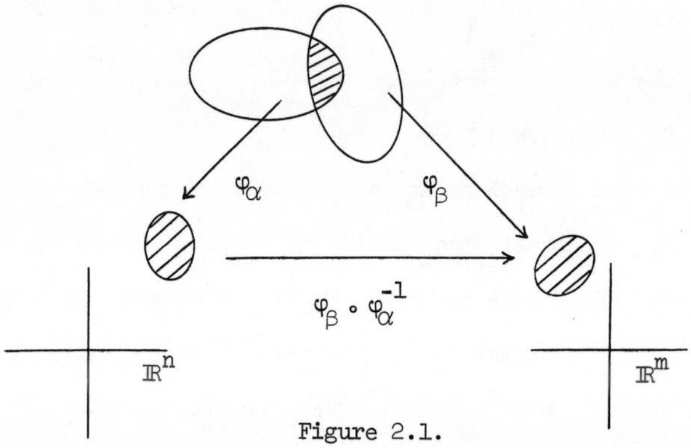

Figure 2.1.

<u>Example 2.3</u>. The circle $S^1 = \{(x,y) \in \mathbb{R}^2 \mid x^2 + y^2 = 1\}$, is a differentiable 1-manifold.

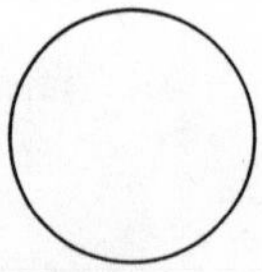

Figure 2.2.

<u>Example 2.4</u>. The sphere $S^2 = \{(x,y,z) \in \mathbb{R}^3 \mid x^2 + y^2 + z^2 = 1\}$, is a differentiable 2-minifold. In this example, a typical chart is $\varphi^{-1}(u,v) = (\cos u \, \cos v, \, \sin u \, \cos v, \, \sin v)$ (see Figure 2.3).

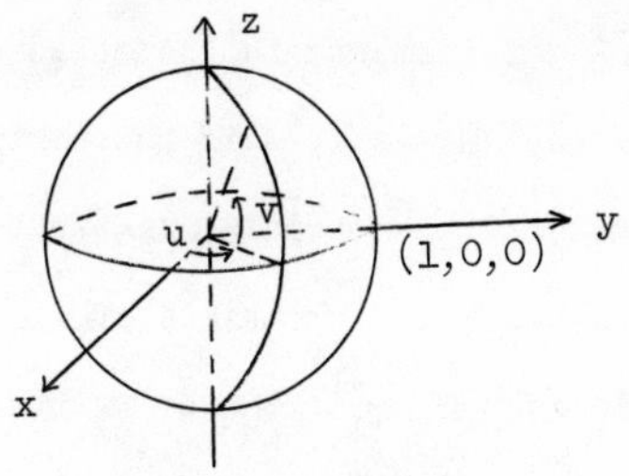

Figure 2.3.

Example 2.5. The torus $T^2 = \{(x,y,z) \in \mathbb{R}^3 \mid (\sqrt{x^2 + y^2} - 2)^2 + z^2 = 1\}$, is a differentiable 2-manifold. In this example, a typical chart is $\varphi^{-1}(u,v) = ((2 + \cos u)\cos v,\ (2 + \cos u)\sin v, \sin u)$ (see Figure 2.4).

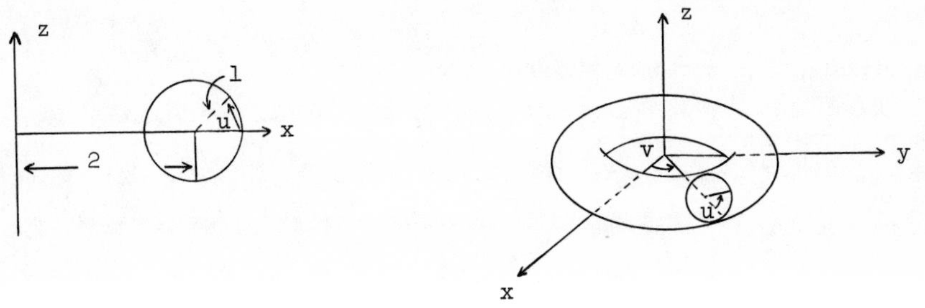

Figure 2.4.

This manifold can also be considered as $\mathbb{R}^2/\mathbb{Z}^2$, where $\mathbb{Z}^2$ is the set of all lattice points with integer coordinates in $\mathbb{R}^2$, with the quotient topology.

Example 2.6. $\mathbb{R}^n$ is an n-dimensional manifold, with an atlas of just one chart given by the identity map.

Example 2.7. The n-torus, $T^n = \mathbb{R}^n/\mathbb{Z}^n$, where $\mathbb{Z}^n$ is the set of all lattice points with integer coordinates, considered as a quotient group of $\mathbb{R}^n$, with φ^{-1} given by the restriction of the projection: $\mathbb{R}^n \to T^n$ to small open sets.

Example 2.8. $G = GL(\mathbb{C},n)$, the set of all complex $n \times n$ invertible matrices, is a $2n^2$-dimensional manifold.

The manifold M is a C^k-, C^∞-, or C^ω (analytic) - manifold if the composition maps $\varphi_\beta \circ \varphi_\alpha^{-1}$ of elements in the atlas is C^k, C^∞, or analytic respectively. In other words, M is C^k if the partial derivatives of $\varphi_\beta \circ \varphi_\alpha^{-1}$ of order less than or equal to k not only exist but also are continuous, M is C^∞ (or smooth) if all partial derivatives of $\varphi_\beta \circ \varphi_\alpha^{-1}$ exist and are continuous, M is analytic if $\varphi_\beta \circ \varphi_\alpha^{-1}$ can be written as a convergent power series.

Let M and N be two differentiable manifolds of dimensions m and n respectively, a map $f\colon M \to N$ is differentiable, C^k, smooth or analytic at $x \in M$ if for each chart (U,φ) at x and each chart (V,ψ) at f(x), the map

$$\psi \circ f \circ \varphi^{-1}\colon \varphi(U \cap f^{-1}(V)) \to \mathbb{R}^m$$

is differentiable, C^k, smooth or analytic respectively.

Let $C^k(M,N)$ be the set of all C^k-mappings from M to N, where k could be ∞ or ω. The C^k-topology in $C^k(M,N)$ is defined as follows. Let ε be any positive real number, then the ε-neighborhood of $f \in C^k(M,N)$ is the set $\{g \in C^k(M,N) \mid \sum_{|\alpha|=1}^{k} |\frac{\partial^{|\alpha|} f}{\partial x^\alpha} - \frac{\partial^{|\alpha|} g}{\partial x^\alpha}| < \varepsilon\}$ where Σ is the sum over all possible α with $\alpha = (\alpha_1,\ldots,\alpha_n)$, $|\alpha| = \sum_{i=1}^{n} \alpha_i$ and

$$\frac{\partial^{|\alpha|} f}{\partial x^\alpha} = \frac{\partial^{|\alpha|} f}{\partial x^{\alpha_1} \cdots \partial x^{\alpha_n}} .$$

Since we are interested in the local situation most of the time, let us review the following definition in euclidean space. Let $f\colon U \to \mathbb{R}^m$ be C^1 at $x_0 \in U$ where U is open in $\mathbb{R}^n$, there is a unique linear transformation $\lambda_{x_0}\colon \mathbb{R}^n \to \mathbb{R}^m$ such that

$$\lim_{h \to 0} \frac{|f(x_0 + h) - f(x_0) - \lambda_{x_0}(h)|}{|h|} = 0 .$$

We denote λ_{x_0} by $df(x_0)$ and call it the Jacobian (or differential) of f at x_0. If we let $(x_1,\ldots,x_n)$ and $(f_1,\ldots,f_m)$ be the local coordinates at x_0 and $f(x_0)$ respectively, then $df(x_0)$ can be represented by a matrix, called the Jacobian matrix of f at x_0, which is denoted by

$$Jf(x_0) = \begin{pmatrix} \frac{\partial f_1}{\partial x_1}, & \ldots, & \frac{\partial f_1}{\partial x_n} \\ & \ldots & \\ \frac{\partial f_m}{\partial x_1}, & \ldots, & \frac{\partial f_m}{\partial x_n} \end{pmatrix} (x_0)$$

Now let M and N be two smooth manifolds and $f\colon M \to N$ be a smooth mapping such that $Jf(x_0)$ (strictly speaking, it should be $J(\psi \circ f \circ \varphi^{-1})(\varphi(x_0))$) is of maximal rank, i.e. $\min\{\dim M, \dim N\}$. Then f is said to be an immersion at x_0 if $\dim M \le \dim N$, a submersion at x_0 if $\dim M \ge \dim N$, a local diffeomorphism at x_0 if $\dim M = \dim N$ and f is a bijective immersion at x_0. Further f is an immersion, submersion or diffeomorphism (globally) if f is an immersion, submersion, or diffeomorphism at each $x_0 \in M$ respectively.

Example 2.9.

$$S^1 \overset{f}{\longmapsto} \mathbb{R}^2$$

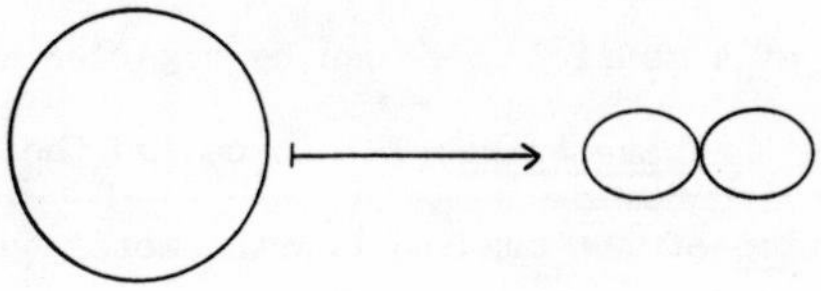

Figure 2.5.

is an immersion.

Example 2.10.

$$\mathbb{R}^1 \overset{g}{\longmapsto} \mathbb{R}^2$$

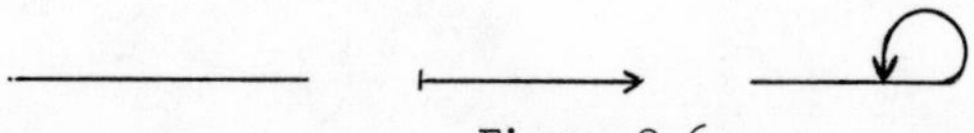

Figure 2.6.

is an immersion.

Example 2.11. S^1 cannot be immersed in $\mathbb{R}^1$.

Example 2.12. Klein bottle can be immersed in $\mathbb{R}^3$.

Example 2.13. Let M be $\mathbb{R}$ with the atlas $\{(\mathbb{R}, \text{identity map})\}$ and N be $\mathbb{R}$ with the atlas $\{(\mathbb{R}, \varphi)\}$ where $\varphi(x) = x^3$, then the map $f: M \to N$ given by $f(x) = x$ is not a diffeomorphism.

For the rest of this section, we will devote ourselves to the definitions of tangent bundle, the tangent space of a smooth manifold and of vector field. These definitions will be mentioned in our later discussions, but not elaborated upon. Therefore we define them in a simple and naive manner. As a consequence, the definitions to follow are handicapped by a possible lack of geometrical motivation.

Let $\{U_\alpha, \varphi_\alpha\}$ be an atlas of a smooth n-manifold M. The tangent bundle of M is an identification space with a projection, denoted by (TM,p) or simply TM. It is defined to be the disjoint union of the spaces $U_\alpha \times \mathbb{R}^n$ over all α with the following identification: whenever $x \in U_\alpha \cap U_\beta$, $(x,v) \in U_\alpha \times \mathbb{R}^n$ is identified with $(x,(\varphi_\beta \circ \varphi_\alpha^{-1})'(v)) \in U_\beta \times \mathbb{R}^n$. The projection $p: TM \to M$ is the usual projection on the first component of each product.

Thus the tangent bundle of a manifold M can be regarded as a triple (TM,M,p), where M is called the base space, TM is called the total space and p is called the projection of the bundle. Furthermore, those mappings $(\varphi_\beta \circ \varphi_\alpha^{-1})'$ are called transition functions. For each point $x \in M$, $p^{-1}(x)$ is called the fibre of the bundle at x or over x. The vector space structure of each fibre can be described by

$$\alpha_1(x,v_1) + \alpha_2(x,v_2) = (x,\alpha_1 v_1 + \alpha_2 v_2) \tag{2.4}$$

where $\alpha_1, \alpha_2 \in \mathbb{R}$ and v_1, v_2 are vectors in $p^{-1}(x)$. The tangent space of M at a point $x \in M$, written as M_x, is defined to be $p^{-1}(x)$ with this linear structure. Each point of M_x is a <u>tangent vector</u> to M at x.

Note that TM can be made into a smooth manifold in an obvious way [66], and p is a smooth map.

The concept of tangent bundle is by no means easy for beginners to understand, as we will see in the first example below, the tangent bundle of a plane is always identified with the plane itself in elementary differential calculus and thus never made explicit. The question is, however, how to glue the pieces together by means of the transition functions, but this is again trivial in this most elementary example.

<u>Example 2.14</u>. The tangent bundle of $\mathbb{R}^n$.

$T\mathbb{R}^n = \{(x,v) \mid x \in \mathbb{R}^n,\ v \in \mathbb{R}^n_x\}$ where $\mathbb{R}^n_x$ is an n-dimensional Euclidean space with the origin at x. The projection map $p: \mathbb{R}^n \times \mathbb{R}^n \to \mathbb{R}$ is defined by $p(x,v) = x$, and with the vector space structure in the fibres defined by (2.4). The fibres are glued together trivially by means of identity function. Thus $T\mathbb{R}^n$ is just a product.

<u>Remark</u>. The tangent bundle of a manifold is called trivial if the total space is (globally) the product of the base space and a vector space with the above mentioned projection and vector space structure in each fibre. If the tangent bundle of M is trivial, then we say M is <u>parallelizable</u>. Thus $\mathbb{R}^n$ is parallelizable.

<u>Example 2.15</u>. The tangent bundle of $S^1 \subset \mathbb{R}^2$. The total space of the tangent bundle of S^1 is again a product. For any $p \in S^1$ we can write

$$p = \alpha(\theta) = (\cos\theta, \sin\theta)$$

where $\alpha: (-\pi,\pi) \to S^1$ is a smooth function. Define

$$v_p = (-\sin\theta, \cos\theta)_p$$

which is indeed $\alpha'(\theta)$. Now if we define

$$f: TS^1 \to S^1 \times \mathbb{R}^1$$

by

$$f(av_p) = (p,a) \quad \text{where} \quad a \in \mathbb{R},$$

it is clear that f is a homeomorphism. This f sends each fibre to $\mathbb{R}^1$ or $f|_{\text{fibre}}$ is a linear isomorphism onto the image.

Thus S^1 is also parallelizable.

To see a non-parallizable manifold, we observe:

Example 2.16. The tangent bundle of $S^2 \subset \mathbb{R}^3$.

Consider the usual manifold structure for S^2. There are two charts given by stereographic projection from the north and south pole

$$P_\epsilon: S^2 - \{(0,0,\epsilon)\} \to \mathbb{R}^2,$$

where $\epsilon = \pm 1$, by sending $(x,y,z) \longmapsto \frac{1}{1 - \epsilon z}(x,y)$. These two charts have transition function

$$L = P_{-1} \circ (P_{+1}| \mathbb{R}^2 - \{(0,0)\})^{-1}: \mathbb{R}^2 - \{(0,0)\} \to \mathbb{R}^2 - \{(0,0)\}.$$

Since $L(x,y) = L^{-1}(x,y) = (\frac{x}{x^2 + y^2}, \frac{y}{x^2 + y^2})$ is differentiable, these two charts determine an atlas for S^2 which gives it the structure of a smooth manifold.

Now

$$dL(x,y) = \frac{1}{(x^2 + y^2)^2} \begin{pmatrix} y^2 - x^2 & -2xy \\ -2xy & x^2 - y^2 \end{pmatrix}$$

gives the identifications that need to be made in order to construct TS^2 from two copies of $\mathbb{R}^2 \times \mathbb{R}^2$, namely identify (x,y,z,w) with

$(L(x,y), dL(x,y)(z,w))$. We can get a somewhat simpler picture of TS^2 by noticing that we get the same space by identifying two copies of $D^2 \times \mathbb{R}^2$ (where D^2 is the unit disk) along $S^1 \times \mathbb{R}^2$ by the restrictions of L and dL to $S^1 = \partial D^2$. If $(x,y) \in S^1$, then $L(x,y) = (x,y)$ and if we write $x = \cos\theta$, $y = \sin\theta$

$$dL(x,y) = \begin{pmatrix} -\cos 2\theta & -\sin 2\theta \\ -\sin 2\theta & \cos 2\theta \end{pmatrix}.$$

The resulting identification then admits the following geometrical description. Let the tangent plane $\{(x,y)\} \times \mathbb{R}^2$ be represented by orthogonal unit vectors $\bar{i}$, $\bar{j}$ parallel to the axes and based at (x,y). Consider $\theta = 0, \pi/4, \pi/2$.

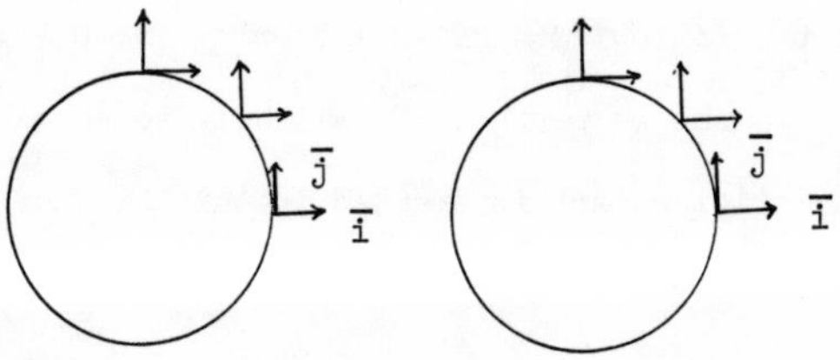

Figure 2.7.

The identifications are those linear maps $\mathbb{R}^2 \to \mathbb{R}^2$ given by multiplication by $\begin{pmatrix} -1 & 0 \\ 0 & 1 \end{pmatrix}$, $\begin{pmatrix} 0 & -1 \\ -1 & 0 \end{pmatrix}$, $\begin{pmatrix} 1 & 0 \\ 0 & -1 \end{pmatrix}$ respectively. So at $\theta = 0$ we identify $\bar{i}$ with $-\bar{i}$, j with $\bar{j}$; at $\theta = \pi/4$, we identify $\bar{i}$ with $-\bar{j}$, j with $-\bar{i}$ and at $\theta = \pi/2$, we identify $\bar{i}$ with $\bar{i}$, $\bar{j}$ with $-\bar{j}$. This is precisely how we would expect the tangent planes to fit together if we think of "bending" the discs into an upper and a lower hemisphere and then gluing them together along their boundaries.

A <u>smooth vector field</u> on a manifold M is a smooth <u>cross section</u> of the tangent bundle, i.e. it is a smooth map $X: M \to TM$ such that $p \circ X = \text{identity}$.

Example 2.17. Consider again the stereographic charts (U,φ_u), (V,φ_v) of S^2 as we described in Example 2.16. Let (u_1,u_2) be the coordinate system in U and (v_1,v_2) be the coordinate system in V. Define

$$X = \begin{cases} (u_1 - u_2)\dfrac{\partial}{\partial u_1} + (u_1 + u_2)\dfrac{\partial}{\partial u_2} & \text{in } U \\ (-v_1 - v_2)\dfrac{\partial}{\partial v_1} + (v_1 - v_2)\dfrac{\partial}{\partial v_2} & \text{in } V \end{cases} .$$

It is not difficult to check that in the intersection of U and V, the two defini equations are the same and hence X provides a globally defined map from S^2 to TS^2.

3. Genericity

To many scientists the "genericity" problem has always been interesting. What we are looking for is the following: given a mapping $f: U \to \mathbb{R}^m$, where U is an open set in $\mathbb{R}^n$, how can we perturb f slightly to obtain a nicer and simpler mapping? Mathematicians are interested in this kind of questions as well. We will state the following two famous theorems to demonstrate the problem of "genericity."

Theorem 3.1. (Classical Weierstrass Approximation Theorem). The set of all polynomials in one variable is dense in the set of all continuous real valued functions defined on the interval $[a,b]$ in the uniform norm, i.e. for every continuous function $f: [a,b] \to \mathbb{R}$, and for every $\epsilon > 0$, there is a polynomial $P(x)$ such that

$$|f(x) - P(x)| < \epsilon$$

for all $x \in [a,b]$.

Theorem 3.2(a). (Whitney [91]) Let M, N be smooth manifolds with $\dim N \geq 2(\dim M)$. Then $I_m(M,N)$ = the set of all immersions from M to N

is an open dense subset of $C^\infty(M,N)$ in the C^∞-topology.

These are two useful theorems. However, from the topological point of view they are not very satisfactory since the maps and their corresponding approximations (perturbed maps) could have quite different topological types. This statement can be illustrated by the following examples.

<u>Example 3.1</u>. (Relating to Theorem 3.1) Let

$$f(x) = \begin{cases} e^{-\frac{1}{x^2}} & x \neq 0 \\ 0 & x = 0 \end{cases}$$

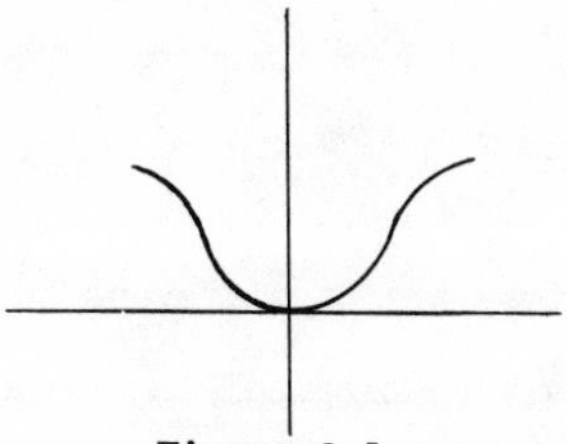

Figure 3.1.

In this case $f^{-1}(0) = \{0\}$. Let $P(x)$ be a polynomial close enough (in the C^0-topology) to f and looking like the curve as indicated in Figure 3.2

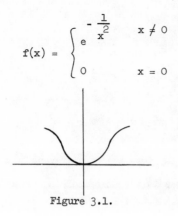

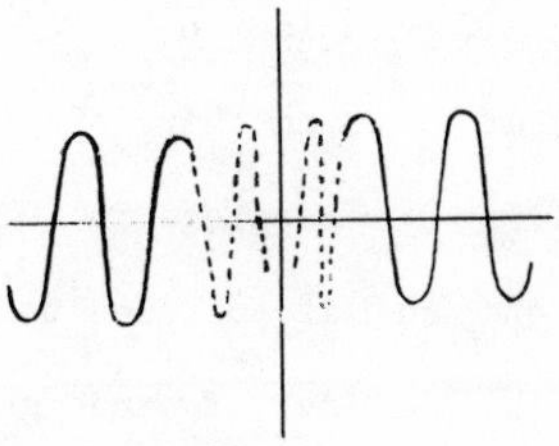

Figure 3.2.

in a neighborhood of 0. It is quite clear that $P^{-1}(0)$ and $f^{-1}(0)$ are not homeomorphic in a neighborhood of 0. We will see that f and P have different topological types and thus f is not stable according to the definition in the next section.

<u>Example 3.2</u>. (Relating to Theorem 3.2(a)) Let $g\colon S^1 \to \mathbb{R}^2$ be defined

as indicated in the diagram

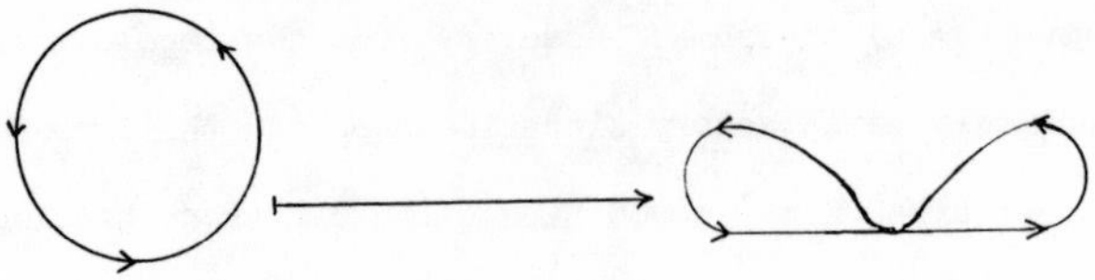

Figure 3.3.

We could pertrub g into either the immersions

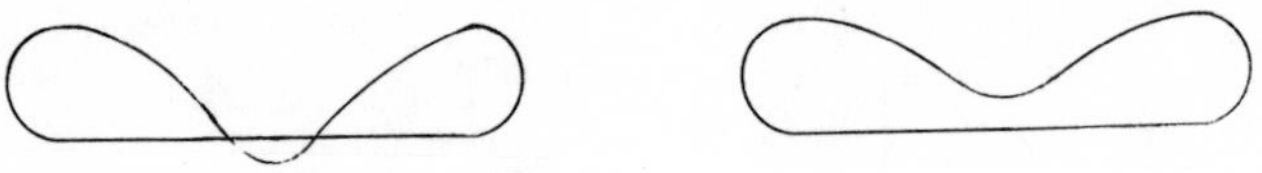

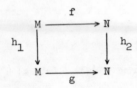

Figure 3.4.

They are not homeomorphic (that is, they are of different topological type), since the self-intersection number is a topological invariance. Thus, g is not stable.

4. Stability

The reason for saying that from a topological point of view, Theorems 3.1 and 3.2(a) are not satisfactory is that we do not have stability although we do have genericity. To define stability for mappings in $C^k(M,N)$, according to the definition in part 2 of this chapter, it suffices to define an equivalence relation in $C^k(M,N)$. Two mappings f and g in $C^k(M,N)$ are equivalent (or strictly speaking C^k-equivalent), $f \sim g$, if there exist C^k-diffeomorphisms (in case $k = 0$, we mean homeomorphisms) $h_1\colon M \to M$ and $h_2\colon N \to N$ such that the diagram

$$\begin{array}{ccc} M & \xrightarrow{\ f\ } & N \\ {\scriptstyle h_1}\big\downarrow & & \big\downarrow{\scriptstyle h_2} \\ M & \xrightarrow[\ g\ ]{} & N \end{array}$$

commutes. Thus f is stable (in the global sense) if the equivalence class of f contains a neighborhood of f in the C^k-topology. This is a rather strong requirement. In many occasions we are only interested in <u>local</u> stability. Let us consider the following example first.

Consider Example 1.1 again, i.e. the map $f(x) = x^2$. Let $g: \mathbb{R} \to \mathbb{R}$ be defined by $g(x) = \epsilon x + x^2$ where ϵ small. Then $g(x) = (x + \frac{1}{2}\epsilon)^2 - \frac{1}{4}\epsilon^2$, or $g(x) + \frac{1}{4}\epsilon^2 = (x + \frac{1}{2}\epsilon)^2$. Set $h_1(x) = x - \frac{1}{2}\epsilon$ and $h_2(y) = y - \frac{1}{4}\epsilon^2$ so that $h_1: (\mathbb{R}, 0) \to (\mathbb{R}, -\frac{1}{2}\epsilon)$, $h_2: (\mathbb{R}, 0) \to (\mathbb{R}, -\frac{1}{4}\epsilon^2)$.

It is clear that

$$g \circ h_1(x) = g(x - \tfrac{1}{2}\epsilon) = x^2 - \tfrac{1}{4}\epsilon^2 = h_2 \circ f(x),$$

i.e., the diagram

$$\begin{array}{ccc} (\mathbb{R}^1, 0) & \xrightarrow{\quad f \quad} & (\mathbb{R}^1, 0) \\ \Big\downarrow h_1 & & \Big\downarrow h_2 \\ (\mathbb{R}^1, -\frac{1}{2}\epsilon) & \xrightarrow[\quad g \quad]{} & (\mathbb{R}^1, -\frac{1}{4}\epsilon^2) \end{array}$$

commutes. This yields a special case of the following definition.

<u>Definition of Local Stability</u>. A C^k-mapping $f: \mathbb{R}^n \to \mathbb{R}^m$ is <u>stable at a point</u> p if there is a neighborhood U of p with the following property: for any neighborhood U' of p with $U' \subset U$, and for any small C^k-perturbation g of f, there is a point $p' \in U'$ and a commutative diagram

$$\begin{array}{ccc} (\mathbb{R}^n, p) & \xrightarrow{\quad f \quad} & (\mathbb{R}^m, f(p)) \\ \Big\downarrow h_1 & & \Big\downarrow h_2 \\ (\mathbb{R}^n, p') & \xrightarrow{\quad g \quad} & (\mathbb{R}^m, g(p')) \end{array}$$

where h_1, h_2 are local C^k-diffeomorphisms at p and $f(p)$ respectively.

Remark. The perturbation map g is required to be C^k. For instance, if we were permitted to perturb the C^2-map $f(x) = x^2$ by a C^1-perturbation $\epsilon x^{3/2}$, then f would not appear to be local stable. However, this kind of artificial difficulty will be eliminated when we come to consider k-jets, which will be defined in section 2 of Chapter 2.

Now we can explain why $f(x) = x^3$ is not stable, which has been stated in section 1. Consider the two perturbed maps,

$$f_1(x) = x^3 - \epsilon x$$

$$f_2(x) = x^3 + \epsilon x ,$$

which may be taken arbitrarily close to f, by taking $\epsilon > 0$ sufficiently small. If they were equivalent, then there would exist local C^k-diffeomorphisms h_1 and h_2 such that

$$f_2 \circ h_1 = h_2 \circ f_1 .$$

Thus,

$$f_2'(h_1(x) \circ h_1'(x) = h_2'(f_1(x)) \circ f_1'(x) ,$$

where h_1' and h_2' will never be zero. Let us regard f as C^2-map. Then $f_2'(h_1(x))$ is zero if $f_1'(x)$ is zero and this is a contradiction. Hence f_1 is not equivalent to f_2, or f is not stable.

Next let us state another part of Whitney's theorem [91] (see Theorem 3.2(a)).

Theorem 4.1 (or 3.2(b)). Let M and N be smooth manifolds with $\dim N \geq 2(\dim M) + 1$. Then the set of all one-to-one immersions from M to N is a residual set, hence dense in $C^\infty(M,N)$. It is open in case M is compact.

The reason we separate Whitney's theorem into two sections may be explained by the following theorem due to J. Mather [20,42]. Basically the reason is that Mather proved stability for mappings under the conditions of Theorem 4.1.

Theorem 4.2. Let M and N be smooth manifolds with M compact, and let $f: M \to N$ be a one-to-one immersion. Then f is stable. If further $\dim N \geq 2(\dim M) + 1$, then f is a one-to-one immersion if and only if it is stable.

The same technique used in proving Theorem 4.2 may also be used to prove:

Theorem 4.3. Let f be a submersion from a manifold M to a manifold Y. Then f is stable.

We would refer the reader to [20,42] (especially [20, p. 72-81]) for the proofs of these statements. The basic ingredient in the proofs of these theorems is Mather's criterion for stability of a mapping, namely that infinitesmal stability implies stability. We will not define the notion of infinitesmal stability in this book, although it is a very important and interesting one. We remark that Theorems 4.1, 4.2 and 4.3 are global results in stability theory, resulting from local hypothesis, and this is the reason for the difficulty in proving them.

Locally, results of the same type - the Implicit Function Theorem and its corollaries - are well known. We will provide them here for later reference.

Theorem 4.4. (Inverse Function Theorem) Let $U \subset \mathbb{R}^n$ be an open set and let $x_0 \in U$. Suppose that $f: U \to \mathbb{R}^n$ is C^k $(k \geq 1)$ and $df(x_0)$ has rank n. Then there is an open set U' containing x_0 and an open set V containing $f(x_0)$ such that $f: U' \to V$ has a C^k inverse $f^{-1}: V \to U'$.

For the proof of this theorem, we refer the readers to [66, p. 35].

Theorem 4.5. (Implicit Function Theorem) Let $U \subset \mathbb{R}^n$, $V \subset \mathbb{R}^m$ be open sets and let $(x_0,y_0) \in U \times V$. Suppose that $f: U \times V \to \mathbb{R}^m$ is C^k

$(k \geq 1)$ and let $(f_1,\ldots,f_m)$ be the component of f. Suppose that $f(x_0,y_0) = 0 \in \mathbb{R}^m$. Let the $m \times m$ matrix M, where

$$M = \begin{pmatrix} \dfrac{\partial f_1}{\partial x_{n+1}}, & \ldots, & \dfrac{\partial f_1}{\partial x_{n+m}} \\ & \ldots & \\ \dfrac{\partial f_m}{\partial x_{n+1}}, & \ldots, & \dfrac{\partial f_m}{\partial x_{n+m}} \end{pmatrix}$$

be non-singular at (x_0,y_0) (i.e., $\det M(x_0,y_0) \neq 0$).

Then there exists open sets $U' \subset U$, $V' \subset V$ containing x_0, y_0 respectively with the following property: for any $x \in U'$, there is a unique $g(x)$ in V' such that $f(x,g(x)) = 0$. Moreover, the function g is also C^k.

<u>Proof</u>: Define $F\colon \mathbb{R}^n \times \mathbb{R}^m \to \mathbb{R}^n \times \mathbb{R}^m$ by $F(x,y) = (x,f(x,y))$. Then

$$dF(x_0,y_0) = \left(\begin{array}{c|c} I_n & 0 \\ \hline * & M(x_0,y_0) \end{array}\right),$$

where I_n is the identity matrix of $\mathbb{R}^n$, has rank $n+m$. By Theorem 4.4 there is an open set $W \subset \mathbb{R}^n \times \mathbb{R}^m$ containing $F(x_0,y_0) = (x_0,0)$ and an open set in $\mathbb{R}^n \times \mathbb{R}^m$ containing (x_0,y_0), which we may take to be of the form $U_1' \times U_2'$ such that $F\colon U_1' \times U_2' \to W$ has a C^k inverse $\tau\colon W \to U_1' \times U_2'$. Clearly τ is of the form $\tau(x,y) = (x,\phi(x,y))$ for some C^k map ϕ (since F is of this form). Let $p_2\colon \mathbb{R}^m \to \mathbb{R}^n$ be defined by $p_2(x,y) = y$; then $p_2 \circ F = f$. Therefore

$$\begin{aligned} f(x,\phi(x,y)) &= f\,\tau(x,y) = p_2 F\tau(x,y) \\ &= p_2(x,y) = y\,. \end{aligned}$$

Thus $f(x,\phi(x,0)) = 0$, in other words we can define $g(x) = \phi(x,0)$, where

$U' \times \{0\} \subseteq W$ and $U_2' = V'$.

<u>Corollary 4.6</u>. Let $f: U \times V \to \mathbb{R}^m$ be C^1 such that $df(x,y)$ has rank m whenever $f(x,y) = 0$. Then $f^{-1}(0)$ is an n-dimensional differentiable manifold in $\mathbb{R}^{n+m}$.

The proof is trivial since at each $(x,y) \in f^{-1}(0)$, (U',φ) is a chart at (x,y), where $\varphi(x) = (x,g(x))$.

<u>Corollary 4.7</u>. Let $f: U \to \mathbb{R}^m$, where $U \subset \mathbb{R}^n$, be a C^k-immersion at $x_0 \in U$. Then f is (locally) stable at x_0. In fact, there are open sets $U' \subset U$, $V \subset \mathbb{R}^m$ with $x_0 \in U'$, $f(U') \subset V$, and there is a C^k-diffeomorphism (or coordinate change) $g: V \to \mathbb{R}^m$ of V onto its image, such that

$$g \circ f(x_1,\ldots,x_n) = (x_1,\ldots,x_n,0,\ldots,0).$$

In other words, we have a normal form for a local immersion.

<u>Proof</u>: Let us write $\mathbb{R}^m = \mathbb{R}^n \times \mathbb{R}^{m-n}$ and write any point in $\mathbb{R}^m$ as (x,y) where $x = (x_1,\ldots,x_n)$, $y = (y_1,\ldots,y_{m-n})$. Consider f as a function $f: U \to \mathbb{R}^n \times \mathbb{R}^{m-n}$, where $U \subset \mathbb{R}^n$, so that f can be written as $f = (f_1,\bar{0}) + (\bar{0},f_2)$ where $f_1: U \to \mathbb{R}^n$ and $f_2: U \to \mathbb{R}^{m-n}$. Since f is an immersion, i.e. $df(x_0)$ has rank m, there is an $n \times n$ minor of $Jf(x_0)$ which is non-singular. We can assume the non-singular $n \times n$ minor of $Jf(x_0)$ is given by the first n-columns. Since if this is not the case, a permutation (which is C^k) of column vectors will do the trick. In other words, we are assuming $df_1(x_0)$ has rank n.

Now, we may construct $F: U \times \mathbb{R}^{m-n} \longrightarrow \mathbb{R}^n \times \mathbb{R}^{m-n}$ by sending $(x,y) \to f(x) + (\bar{0},y)$ where x and $\bar{0}$ are in $\mathbb{R}^n$ and $y \in \mathbb{R}^{m-n}$. Then

$$dF(x_0,y) = \left(\begin{array}{c|c} df_1(x_0) & 0 \\ \hline * & I_{m-n} \end{array}\right)$$

has rank m. By the Inverse Function Theorem (Theorem 4.4), there exists an inverse g to F on a neighborhood $(x_0,\bar{0})$ such that

$$g \circ f(x) = g \circ F(x,\bar{0}) = (x,\bar{0}) .$$

Corollary 4.8. Let $f: \mathbb{R}^n \to \mathbb{R}^m$, $f = (f_1,\ldots,f_m)$, be a C^k-submersion at $x_0 \in U$. Then f is (locally) stable at x_0. In fact, there is an open set $U' \subset U$ and a C^k-diffeomorphism $h: U' \to \mathbb{R}^n$ of U' onto its image such that

$$f \circ h(x_1,\ldots,x_n) = (x_{n-m+1},\ldots,x_n) .$$

Note that this is the normal form for a local submersion.

Proof: Let us write $\mathbb{R}^n \xrightarrow{f} \mathbb{R}^{n-m} \times \mathbb{R}^m$. Just as we indicated in Corollary 4.7, let us assume the matrix $(\frac{\partial f_i}{\partial x_j})$, $1 \leq i, j \leq m$, is non-singular. Then we are precisely in the situation considered in the proof of the Implicit Function Theorem (Theorem 4.5), and thus there is a C^k-diffeomorphism h such that

$$g \circ h(x_1,\ldots,x_n) = (x_{n-m+1},\ldots,x_n) .$$

5. Singularities

In general, we cannot expect the maps are either immersions or submersions. The following definition arises naturally.

Definition 5.1. Let f be a differentiable mapping from M to N, where M and N are differentiable manifolds. A point $x_0 \in M$ is a singular point of f if rank $df(x_0) < \min\{\dim M, \dim N\}$. Otherwise, x_0 is called a regular point of f.

Example 5.1. $f(x) = x$ has no singular point.

Example 5.2. 0 is a singular point of $f(x) = x^2$ where $x \in \mathbb{R}^1$.

<u>Example 5.3</u>. 0 is a singular point of $f(x) = x^3$ where $x \in \mathbb{R}^1$.

<u>Example 5.4</u>. Let $f\colon \mathbb{R}^2 \to \mathbb{R}^1$ be defined by

$$f(x,y) = xy - x^3 .$$

With the natural coordinates,

$$Jf(x,y) = (y - 3x^2, x) .$$

$(0,0)$ is the only singular point of the map.

<u>Example 5.5</u>. Let $g\colon \mathbb{R}^2 \to \mathbb{R}^2$

$$(x,y) \to (u,v)$$

where

$$\begin{cases} u = xy - x^3 \\ v = y \end{cases} .$$

With the coordinate system (x,y) and (u,v),

$$Jg(x,y) = \begin{pmatrix} y - 3x^2 & x \\ 0 & 1 \end{pmatrix} .$$

Thus, the set of singular points of this map is $\{(x,y) \in \mathbb{R}^2 \mid y = 3x^2\}$.

It is clear that a point to be a singularity of a mapping is a local property. We will mostly focus attention on mappings: $\mathbb{R}^n \to \mathbb{R}^m$ which have a singularity at $\bar{0}$.

Now we turn our attention to the Morse lemma.

<u>Definition 5.2</u>. Let $f\colon \mathbb{R}^n \to \mathbb{R}^1$ be a smooth function. A point $\bar{x}_0$ in $\mathbb{R}^n$ is a <u>non-degenerate critical point</u> if $\bar{x}_0$ is a singular point of f and the Hessian, which is the determinant of the matrix

$$\left(\frac{\partial^2 f}{\partial x_i \partial x_j}\right), \quad 1 \leq i, j \leq n$$

evaluated at $\bar{x}_0$, is non-zero.

Definition 5.3. A smooth function $f: \mathbb{R}^n \to \mathbb{R}$ is a Morse function is all its singular points are non-degenerate critical points. (The domain $\mathbb{R}^n$ of f could be replaced in this definition by any differentiable manifold M.)

Theorem 5.1. (The Morse Lemma) [51]

(1) Morse functions on $\mathbb{R}^n$ are (locally) stable at their non-degenerate critical points.

(2) The set of all Morse functions is dense in $C^\infty(\mathbb{R}^n, \mathbb{R}^1)$.

(3) Let $\bar{x}_0$ be a non-degenerate critical point of a function $f: \mathbb{R}^n \to \mathbb{R}$. Then there is a number λ such that, for a suitable change of coordinates, $f(\bar{x})$ can be written as

$$f(\bar{x}) = x_1^2 + \dots + x_\lambda^2 - x_{\lambda+1}^2 - \dots - x_n^2$$

in a neighborhood of $\bar{x}_0$. (This is the normal form for a Morse function at a singular point.)

As a consequence of Mather's criterion for stability (i.e. that infinitesmal stability implies stability), the global version of the Morse lemma is also true.

Theorem 5.2. Let M be a smooth n-dimensional manifold.

(1) $f: M \to \mathbb{R}^1$ is stable (globally) if and only if f is a Morse function on M and the critical values are distinct. (i.e. if x_1, x_2 are singular points of f then $f(x_1) \neq f(x_2)$; such a function f has been referred as a nice function.

(2) The set of all Morse functions is open and dense in $C^\infty(M, \mathbb{R}^1)$.

Finally, let us state Whitney's theorem in dimension two [87].

Theorem 5.3.

(1) The mapping $f: U \to \mathbb{R}^2$ is stable at $\bar{x}_0 \in U \subset \mathbb{R}^2$ if and only if it is equivalent in some neighborhood of $\bar{x}_0$ to one of the three mappings:

(i) $u = x,\ v = y$ (regular point),

(ii) $u = x^2,\ v = y$ (fold point),

(iii) $u = xy - x^3,\ v = y$ (cusp point)

each mapping a neighborhood of $(0,0)$ in the (x,y) plane into a neighborhood of $(0,0)$ in (u,v) plane.

(2) The stable mappings $f: X \to \mathbb{R}^2$ of a compact 2-dimensional surface into the plane form an everywhere dense set in the space of all smooth mappings.

(3) The smooth mapping $f: X \to \mathbb{R}^2$ is stable if and only if the following two conditions are satisfied:

(a) The mapping is stable at every point in X.

(b) The images of folds intersect only pair-wise and at non-zero angles, whereas images of folds do not intersect images of cusps.

This theorem will be discussed in detail in the next chapter. One final remark is that the condition that M be compact in statement (2) of this theorem enables us to use Thom's Transversality Theorem (which will be discussed later).

CHAPTER 2

ON SINGULARITIES OF MAPPINGS FROM THE PLANE INTO THE PLANE

1. Introduction

The emphasis in this chapter will be on the key paper by H. Whitney published in 1955, which should be regarded as a landmark in the development of the theory of singularities of mappings. In this paper, Whitney not only proved the remarkable theorem which I stated in the last chapter but also did something else which turned out to be very important in later developments in singularity theory. Namely, he found that information about the behavior of differentiable functions is contained in the values of its derivatives, and he was also able to formulate a very useful concept in this regard. Thus to extract information about a map f it makes sense to consider as a separate mathematical object a certain space, which will be called a Jet Space, which possesses as its points the values of the rth order derivatives of a function, for some r. It was also in this paper that Whitney observed that the non-degeneracy criterion which the Morse function must satisfy is merely the condition that, when the first order partials all vanish, the second order partials will not lie in a certain proper algebraic subset of the jet space, referred to as the "bad" set, and defined by the vanishing of a finite set of polynomials.

We will also demonstrate the essential steps in making a coordinate change to obtain the normal form for a mapping from the plane to the plane with fold points only. In order to clarify the "bad" sets and describe Whitney's method and results, we give certain definitions which will also help us to simplify the proofs of the theorems mentioned above.

2. Jet Space

Given a local C^s-mapping $f = (f_1,\dots,f_p)\colon \mathbb{R}^n \to \mathbb{R}^p$, with $f(\bar{0}) = \bar{0}$, we may expand each f_i in a Taylor expansion about the origin. If we omit all the terms of degree $\geq r + 1$ $(r < s)$, what remains is a p-tuple of polynomials of degree r, which approximates f. Such a p-tuple is called an r-jet. This is the intuitive definition of an r-jet since it depends on the choice of a coordinate system. We now give the coordinate-free definition.

Definition 2.1. Let $C^s(n,p)$ be the set of all s-times continuously differentiable mappings, $f = (f_1,\dots,f_p)\colon \mathbb{R}^n \to \mathbb{R}^p$, with $f(\bar{0}) = \bar{0}$. We call $f,g \in C^s(n,p)$ equivalent of order r at $\bar{0}$, if at $\bar{0} \in \mathbb{R}^n$, their formal Taylor expansions up to and including the terms of degree $\leq r$ are identical. The r-jet of f, denoted by $j^r(f)$ (it may also be denoted $J^r(f)(\bar{0})$) is the equivalence class of f; and f is called a realization of the jet $j^r(f)$. For further study of jets and their examples, we refer the reader to Chapter 6.

The set of all r-jets is denoted by $J^r(n,p)$. Taking the values of the partial derivatives at $\bar{0}$ as the coordinates of a jet, $J^r(n,p)$ becomes a Euclidean space. Thus an r-jet can either be realized by a set of truncated polynomials of degree $\leq r$ or by an N-tuple, $\alpha = (\alpha_1,\dots,\alpha_N)$ for some N, in the Euclidean space $J^r(n,p)$.

For a C^s-mapping $f\colon U \to \mathbb{R}^p$, where U is an open subset of $\mathbb{R}^n$, $f(\bar{0}) = \bar{0}$, the r-extension of f

$$J^r(f)\colon U \to J^r(n,p)$$

is defined as follows: for $x_0 \in U$, translate the origins of $\mathbb{R}^n$ and $\mathbb{R}^p$ to x_0 and $f(x_0)$ respectively, then $J^r(f)(x_0)$ is the Taylor expansion of f at x_0 up to and including the terms of degree $\leq r$.

Example 2.1. Let $f\colon \mathbb{R} \to \mathbb{R}$ be defined by

$$f(x) = x^3 . \tag{2.1}$$

Near x_0 in $\mathbb{R}$,

$$\begin{aligned} f(x) &= f(x - x_0 + x_0) = (x - x_0 + x_0)^3 \\ &= x_0^3 + 3x_0^2(x - x_0) + 3x_0(x - x_0)^2 + (x - x_0)^3 . \end{aligned}$$

By definition,

$$J^3 f(x_0) = 3x_0^2 x + 3x_0 x^2 + x^3 \in J^3(1,1) ,$$

where $J^3(1,1)$ can be identified with $\mathbb{R}^3$ with the correspondence

$$ax + bx^2 + cx^3 \leftrightarrow (a,b,c) .$$

Thus, it is clear that $J^3 f$, where $f(x) = x^3$, maps $\mathbb{R}^1$ onto the parabola $a = 3x_0^2$, $b = 3x_0$, $c = 1$ in $\mathbb{R}^3$. Diagrammatically, we have

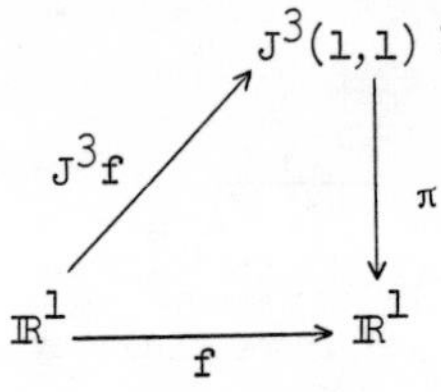

,

where π is the projection onto c-axis. This example will be studied further later on.

Remark. The above definition can be generalized from $J^r(n,p)$ to $J^r(M,Y)$ where M, Y are smooth manifolds. For the generalized definition, we refer the reader to [20, p. 37; 34, p. 23] .

3. Transversality

Transversality is a very important and profound idea in the study of singularity theory. Thom's transversality theorem provides a complete and thoroughly successful resolution and clarification of one of the basic ideas in singularity theory, namely genericity. The applications of Thom's theorem will be seen in section 4 of this chapter as well as section 4 in Chapter 4.

Intuitively, transversality can be illustrated by the local intersection of two curves P, Q differentiably embedded in $\mathbb{R}^2$ as follows:

Case 1: $P = \{(x,y) \mid x = 0\}$.

$Q = \{(x,y) \mid x - a = y^2 \text{ where } a > 0\}$.

Figure 3.1.

Case 2: P same as above.

$Q = \{(x,y) \mid x = y^2\}$.

Figure 3.2.

Case 3: P same as above.

$Q = \{(x,y) \mid x + a = y^2 \text{ where } a > 0\}$.

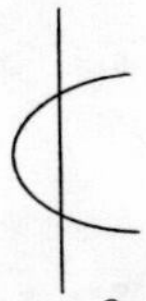

Figure 3.3.

Case 4: P same as above.

$Q = \{(x,y) \mid x = y^4\}$.

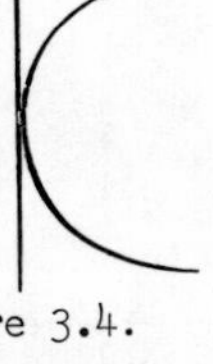

Figure 3.4.

In Cases 1 and 3, the two curves P, Q intersect transversally since a small perturbation of either one of the maps would not change the type of intersection. However, this is no longer true for Cases 2 and 4 because a small deformation in Case 2 or 4 will change their type of intersection into Case 1 or Case 3 or even in Case 4, the type of intersection can be changed into the following type of intersection.

Figure 3.5.

Definition 3.1. Let M and Y be smooth manifolds and let $f\colon M \to Y$ be a smooth mapping. Let S be a submanifold of Y and let $x \in M$. Then f intersect S transversally at x (in short, $f \pitchfork S$ at x) if either

(1) $f(x) \notin S$ or

(2) $f(x) \in S$ and $T_{f(x)}S + (df)_x(T_xM) = T_{f(x)}Y$,

where T_xM is the tangent space to M at x.

Definition 3.2. Let M, Y, S and f be as above. Let A be a subset of M, then $f \pitchfork S$ on A if $f \pitchfork S$ at each $x \in A$. Moreover $f \pitchfork S$ if $f \pitchfork S$ on M.

Example 3.1. Let $M = \mathbb{R}$, $Y = \mathbb{R}^2$, S = x-axis in $\mathbb{R}^2$, and let $f\colon \mathbb{R} \to \mathbb{R}^2$ be defined by

$$f(x) = (x, (x + 1)^2).$$

Then $f \pitchfork S$.

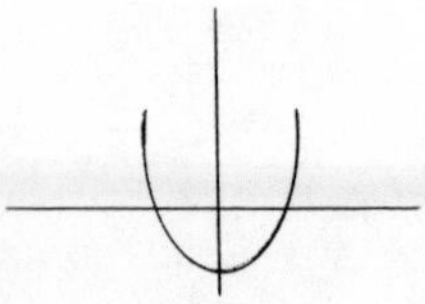

Figure 3.6.

Example 3.2. Let M, Y, S be as above, Define $f\colon \mathbb{R} \to \mathbb{R}^2$ to be

$$f(x) = (x,x^2) .$$

Then f does not intersect transversally with S at (0,0) .

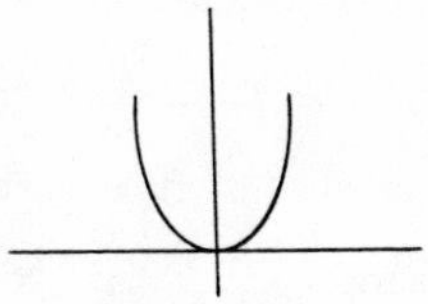

Figure 3.7.

Example 3.3. Let M, Y, S be as above. Define $f\colon \mathbb{R} \to \mathbb{R}^2$ to be

$$f(x) = (x,x^3) .$$

Then f does not intersect transversally with S at (0,0) .

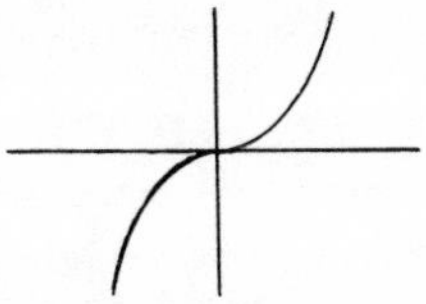

Figure 3.8.

Definition 3.3. Let M and S be submanifolds of an ambient smooth manifold Y . Then $M \pitchfork S$ if either $M \cap S = \emptyset$ or $T_xM + T_xS = T_xY$ for all $x \in M \cap S$. Thus $M \pitchfork S$ precisely when $f \pitchfork S$ for the embedding f of M in Y .

Alternatively we have $M \pitchfork S$ if the sum of the normal spaces of M and S is a direct sum at every $x \in M$. (Convention: if $x \notin S$, the normal space of S at $x = \eta_x S = \{0\}$.)

The equivalence of these two definitions is obvious since (tangent space) + (tangent space) = whole space if and only if their orthogonal complements have zero intersection.

The alternative definition can easily be generalized to define the transversality of a family of submanifolds.

Definition 3.4. A family of submanifolds $\{S_i\}$ of Y is transversally intersected at x if the sum of normal spaces $\eta_x S_i$ is a direct sum.

For our applications, we need instead of simply a submanifold S, a stratified submanifold (or, as Whitney called it, a manifold collection). This definition can be motivated by the following special case. Let L be the space of linear transformations of a vector space V (of dimension n) into a vector space W (of dimension m). Let $F \in L$. The rank of F is the dimension of the image space $F(V)$ in W. Let L_k denote the set of elements in L of rank k. Now

$$L = L_0 \cup \cdots \cup L_\nu$$

where $\nu = \min\{n,m\}$. It is easy to see that each L_k is a manifold, of dimension

$$\dim(L_k) = nm - (n-k)(m-k) .$$

The codimension of L_k in L is denoted by

$$\begin{aligned} \operatorname{codim}(L_k) &= \dim(L) - \dim(L_k) \\ &= nm - [nm - (n-k)(m-k)] \\ &= (n-k)(m-k) . \end{aligned}$$

The expression $L = L_0 \cup \cdots \cup L_\nu$, $\nu = \min\{m,n\}$, is said to be a stratified manifold.

The idea here is that the closure of the point set L_k is a manifold "with singularities" and the set of singularities is exactly $L_0 \cup \cdots \cup L_{k-1}$.

Definition 3.5. A stratified submanifold S of $\dim m$ in Y (in this book, we always regard Y as a sufficiently high dimensional Euclidean space) is a set of disjoint submanifolds $S_0, S_1, \ldots, S_\ell$ such that $m = \max_{1 \le i \le \ell} \{\dim S_i\}$ and each point set $S_0 \cup \cdots \cup S_i$ is closed in Y.

Definition 3.6. Let $S \subset Y$ be a stratified submanifold and let $f: M \to Y$ be a smooth mapping. Then $f \pitchfork S$ if $f \pitchfork S_i$ for each i.

Theorem 3.1. (Thom's Transversality Theorem) Let S be a smooth submanifold (or stratified submanifold) of $J^r(n,p)$. Then, for almost all C^∞-mappings $f: \mathbb{R}^n \to \mathbb{R}^p$, $J^r f(\mathbb{R}^n) \pitchfork S$. If further $\operatorname{codim} S > n$, where $\operatorname{codim} S$ is the codimension of S in $J^r(n,p)$, then $J^r(f)(\mathbb{R}^n) \cap S = \emptyset$ for almost all f.

An important special case occurs when $r = 0$.

Theorem 3.2. Let M and S be two submanifolds of $\mathbb{R}^n$. Then in general $M \pitchfork S$; and, if $\operatorname{codim} S > \dim M$, then $M \cap S = \emptyset$ in general.

This theorem is of course true if S is a stratified submanifold. The proof of this theorem can be found in [45, 58, 81]. We will not go into details here.

4. Morse Lemma--The Genericity Aspect

In this section we elaborate the idea which has been described informally in section 1 and prove the density of Morse functions; namely, given any smooth mapping $f_0: \mathbb{R}^n \to \mathbb{R}^1$ with singular point at $\bar{x}_0 \in \mathbb{R}^n$, we shall find a smooth mapping $f: \mathbb{R}^n \to \mathbb{R}^1$ close enough (in the C^∞-topology) to f_0 with non-degenerate critical point at $\bar{x}_0$.

Consider the mapping $J^2 f: \mathbb{R}^n \to J^2(n,1)$ by sending a point $p \in \mathbb{R}^n$ into $J^2 f(p) = \left(\frac{\partial f}{\partial x_1}, \ldots, \frac{\partial f}{\partial x_n}, \frac{\partial^2 f}{\partial x_1^2}, \frac{\partial^2 f}{\partial x_1 \partial x_2}, \ldots, \frac{\partial^2 f}{\partial x_n^2}\right)_p \in J^2(n,1)$. As we remarked before, we can consider any point $\alpha \in J^2(n,1)$ as an $N = \left(n + \frac{n(n+1)}{2}\right)$ - tuple, $\alpha = (\alpha_1, \ldots, \alpha_n, \alpha_{11}, \alpha_{12}, \ldots, \alpha_{nn})$. Thus $J^2(n,1) \cong \mathbb{R}^N$ where $\cong$ is an isomorphism of vector spaces.

The bad set in $J^2(n,1)$ is the set of points, $\alpha = (\alpha_1, \ldots, \alpha_n, \alpha_{11}, \ldots, \alpha_{nn})$, such that, setting $\alpha_{ij} = \alpha_{ji}$ for $i > j$, $\alpha_1 = \alpha_2 = \ldots = \alpha_n = 0$ and

$$D = D(\alpha) = \begin{vmatrix} \alpha_{11}, \dots, \alpha_{1n} \\ \dots\dots\dots \\ \alpha_{n1}, \dots, \alpha_{nn} \end{vmatrix} = 0 . \qquad (4.1)$$

Let S_i be the subset of S in which rank $D = i$, then $S = S_0 \cup \dots \cup S_{n-1}$. We shall use Whitney's argument [87] to outline below a proof that S is a stratified manifold of $\dim(\frac{n(n+1)}{2} - 1)$. Granted this, we use Thom's Transversality Theorem. Thus, since $\dim S + \dim \mathbb{R}^n = \frac{n(n+1)}{2} - 1 + n < N$, we know that for almost all $f: \mathbb{R}^n \to \mathbb{R}^1$, $J^2 f(\mathbb{R}^n) \cap S = \emptyset$. This implies that for almost all f, $\bar{x}_0$ is non-degenerate in case it is a critical point of f.

To insure that S is a stratified manifold of dimension $(\frac{n(n+1)}{2} - 1)$, we show that S_i is a smooth manifold of dimension

$$m_i = \binom{i+1}{2} + i(n-i) = \binom{n+1}{2} - \binom{n-i+1}{2} . \qquad (4.2)$$

Given $\alpha \in S_i$, there are i rows of $D = D(\alpha)$ which are independent. Without loss of generality, we assume that these i rows are $A_1, \dots, A_i$, and that the principal minor with these rows is non-zero. Now take any $\alpha' \in S_i$ near α. Then the rows $A'_1, \dots, A'_i$ are independent, and each other row is dependent on them, and so has a unique expression

$$A'_k = a_{k1} A'_1 + \dots + a_{ki} A'_i \qquad (4.3)$$

where $i + 1 \leq k \leq n$. This gives

$$\alpha'_{kj} = \alpha'_{jk} = a_{k1}\alpha'_{1j} + \dots + a_{kh}\alpha'_{hj} + \dots + a_{ki}\alpha'_{ij} \qquad (4.4)$$

where $1 \leq h \leq i$. They hold for $j = 1, 2, \dots, n$; however only i out of j values will yield independent equations. Hence, we have $i(n - i)$ independent equations. By choosing $\binom{i+1}{2}$ elements α'_{hj} with $h \leq j$ arbitrarily, it is clear that these equations can be solved for that a_{kh}, giving them as fixed analytic functions of the elements of the rows $A'_1, \ldots, A'_i$. Thus S_i is a smooth manifold, and its dimension is the number of independent elements, namely $m_i = \binom{i+1}{2} + i(n-i) = \binom{n+1}{2} - \binom{n-i+1}{2}$.

Next, we show that $S_0 \cup \ldots \cup S_i$ are closed. For any sequence of elements $\alpha^{(1)}, \alpha^{(2)}, \ldots$ of S_i, with a limit α^*, let D^* be the corresponding determinant of α^*, we have rank $(D^*) \leq i$ since α^* is the limit of $\alpha^{(1)}, \alpha^{(2)}, \ldots$ Hence $\alpha^* \in S_q$ with $q \leq i$.

Finally, $\dim S = \max\limits_{0 \leq i \leq n-1} m_i = \binom{n+1}{2} - \binom{n-(n-1)+1}{2} = \binom{n+1}{2} - 1$,

which is what we want to prove.

5. Characterization of Folds and Cusps

We consider the following two examples, which give the normal forms of the fold and the cusp, as a guide to the study of Whitney's characterization of folds and cusps.

Example 5.1. Consider the mapping $f: \mathbb{R}^2 \to \mathbb{R}^2$ such that $f(x,y) = (u,v)$ where $u(x,y) = x^2$, $v(x,y) = y$. Then the Jacobian matrix of f is

$$\begin{pmatrix} 2x & 0 \\ 0 & 1 \end{pmatrix}.$$

We fix the following notation: let J stand for the determinant of the Jacobian matrix of f if there is no confusion to be feared. In this example, we have $J(x,y) = 2x$. By definition, a point where $J = 0$ is called a singular point of f. Thus, every point $(0,y)$ is a singularity of f. Moreover, at any point $(0,y)$ on the y-axis, the cross section of the graph (see Figure 5.1) is the same as Figure 1.1 of Chapter 1 and we know it is stable. For the same

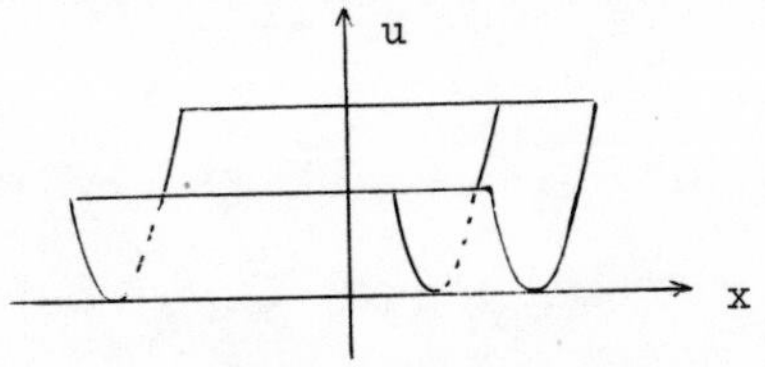

Figure 5.1.

as given in Example 1.1 in section 1 of Chapter 1, any small movement (or perturbation) of the mapping will not change its topological type (the proof will be given in section 6 of this chapter). The y-axis is called the "fold" of f: this notion will be defined and given geometrical interpretation in the latter part of this section. In particular

$$\begin{cases} u = x^2 \\ v = y \end{cases} \tag{5.1}$$

is called the <u>normal form of a fold</u>. Whitney called it the normal form because for any mapping which satisfies the condition of a "fold" at its singular point can be transformed (through changing coordinates) into this particular form (see Theorem 6.2 in this chapter).

<u>Example 5.2</u>. Let $g: \mathbb{R}^2 \to \mathbb{R}^2$ be defined by $g(x,y) = (u,v)$ where

$$\begin{cases} u(x,y) = xy - x^3 \\ v(x,y) = y \end{cases} \tag{5.2}$$

The graph of the function u is as Figure 5.2.

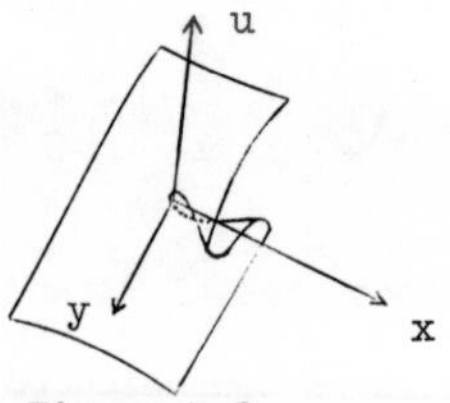

Figure 5.2.

The Jacobian matrix of g is

$$\begin{pmatrix} y - 3x^2 & x \\ 0 & 1 \end{pmatrix},$$

and

$$J(x,y) = y - 3x^2 . \tag{5.3}$$

The singularities of g lie along the parabola $y = 3x^2$ in the (x,y) plane.

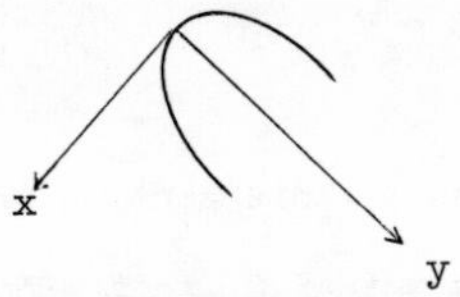

Figure 5.3.

This is called the <u>general fold</u> of this mapping. As we have discussed in Chapter 1 the surface $u(x,y) = x^3 - xy$ is the stable unfolding of the mapping $f(x) = x^3$. The map g is called the <u>normal form of a cusp</u>.

Next we will give some general properties of mappings from $\mathbb{R}^2$ into $\mathbb{R}^2$ (although the notions can be extended to higher dimensions), and consider in particular the directional derivatives.

Let $f: U \to \mathbb{R}^2$ be a smooth mapping where U is open in $\mathbb{R}^2$. Let W be a vector in $\mathbb{R}^2$, $p \in U$, $t \in \mathbb{R}$; then the directional derivative of f at p in the direction W is defined as usual by

$$\nabla_W f(p) = \lim_{t \to 0^+} \frac{1}{t}\{f(p + tW) - f(p)\} . \tag{5.4}$$

It is obvious that $\nabla f(p)$ is a linear mapping from vectors in $\mathbb{R}^2$ to vectors in of $\mathbb{R}^2$. In particular, if e_1, e_2 are the unit vector of a coordinate system (x_1,x_2) at p, then

$$\nabla_{e_i} f(p) = \frac{\partial f}{\partial x_i}(p) . \tag{5.5}$$

Therefore, in general,

$$\nabla_W f(p) = \sum_{i=1}^{2} (w_i \frac{\partial f}{\partial x_i})(p) , \tag{5.6}$$

where $W = (w_1,w_2)$ in this coordinate system. Write (x,y) instead of (x_1,x_2), and $f(x,y) = (u(x,y),v(x,y))$,

$$\nabla_W f(p) = w_1 f_x(p) + w_2 f_y(p) = \begin{pmatrix} w_1 u_x + w_2 u_y \\ w_1 v_x + w_2 v_y \end{pmatrix}(p) . \tag{5.7}$$

We can certainly differentiate again. Let $Z = (z_1, z_2)$ be a vector in $\mathbb{R}^2$.

$$\nabla_Z \nabla_W f(p) = z_1(w_1 f_x + w_2 f_y)_x(p) + z_2(w_1 f_x + w_2 f_y)_y(p) . \tag{5.8}$$

Returning to the two examples in this section let us take a point Q in Figure 5.1 (or more accurately, a point Q in the image of f). Let W be in the direction of the positive x-axis. It is clear that $\nabla_W f(Q)$ approaches zero as Q approaches P , which is $(0,y)$ in other words $\nabla_W f(p) = 0$. This is also clear from the view-point of physics since $f(0,y)$ is a stagnation point, i.e. the velocity vector becomes zero as Q approaches P . Let us remark that the point Q cannot be taken to be a point on the y-axis, in fact Q could be taken any point in the image of f away from the y-axis and if we connect Q to P by any path, then along the tangent direction of this path $\nabla_W f(p)$ approaches zero as Q approaches P . We could do the same thing on Figure 5.2 . In this case, the stagnation points are on the parabola $y = 3x^2$; in other words, taking any point $Q(x,y)$ with $y \neq 3x^2$, and any $P(x,y)$ with $y = 3x^2$, we connect Q to P by any path ϕ , and then $\nabla_W f(p) = 0$ where W is the tangent vector field of the path ϕ (i.e. the function which assigns to any point $\phi(t)$ the tangent vector to ϕ at $\phi(t)$.

Notice that if we take Q on the parabola $y = 3x^2$ and if the path ϕ is a portion of the parabola connection P and Q , then $\nabla_W f(p) \neq 0$ except at $p = (0,0)$. At $(0,0)$, $\nabla_W f(p) = 0$ as Q approaches P along $y = 3x^2$.

All these remarks are quite clear by looking at the Figures 5.1 and 5.2 and they lead to the following definitions.

Let $\phi\colon \mathbb{R}^1 \to \mathbb{R}^2$, $t \to \phi(t) = (x,y)$, be a C^2-parametrized curve in $\mathbb{R}^2$. The tangent vector of ϕ is $\frac{d\phi}{dt}(t) = \nabla_e \phi(t)$ where e is the unit vector in the positive direction of $\mathbb{R}^1$. Let W be a non-zero C^1-vector field in $\mathbb{R}^2$ such that

$$W(\phi(t)) = \frac{d}{dt}\phi(t) . \tag{5.9}$$

Then, by the chain rule, we have

$$\frac{d}{dt}(f \circ \phi)(t) = \frac{\partial f}{\partial x}(\phi(t))\frac{\partial \phi}{\partial t}(t) + \frac{\partial f}{\partial y}(\phi(t))\frac{d\phi}{dt}(t)$$
$$= \frac{\partial f}{\partial x}(p) \cdot W + \frac{\partial f}{\partial y}(p) \cdot W = \nabla_W f(p) \tag{5.10}$$

at $p = \phi(t)$. Similarly,

$$\frac{d^2}{dt^2}(f \circ \phi)(t) = \nabla_W \nabla_W f(p) \tag{5.11}$$

at $p = \phi(t)$. Since

$$\nabla_W f(p) = w_1(u_x, v_x) + w_2(u_y, v_y) , \tag{5.12}$$

it is clear that $\nabla_W f(p) = 0$ if and only if $u_x v_y - v_x u_y = 0$ for any $W \neq 0$. Thus p is a regular point of f if and only if $\nabla_W f(p) \neq 0$ whenever $W \neq 0$, otherwise p is a singular point of f.

Definition 5.1. Let f be a C^2-function. A point p is a good point if either $J(p) \neq 0$ or $\nabla J(p) \neq 0$ (i.e. at least one of J_x and J_y is nonzero). We say that f is good if every point in dom f is good.

Remark 5.1. In case $u_x = u_y = v_x = v_y = 0$, then $\nabla J(p) = 0$, so p cannot be good. Thus if f is good, then the image space of the linear map $\nabla f(p)$ from the 2-dimensional vector space $\mathbb{R}^2$ into $\mathbb{R}^2$ is of dimension two if p is regular and is of dimension 1 if p is singular points, respectively.

Lemma 5.1. Let f be good in $U \subset \mathbb{R}^2$. Then the singular points of f form a smooth curve in U.

Proof: Let $p = (x_0, y_0)$ be a singular point of f, i.e. $J(x_0, y_0) = 0$. Since f is good, $\nabla J(p) \neq 0$. By the Implicit Function Theorem (Theorem 4.5 of Chapter 1), there exists a neighborhood A of x_0, a neighborhood B of y_0 and a smooth function $\gamma: A \to B$, such that $\gamma(x_0) = y_0$ and $J(x, \gamma(x)) = 0$ for all $x \in A$. Therefore the solutions of $J = 0$ near a good point of f lie on a smooth curve C, which is called the general fold of f.

Definition 5.2. Let $\phi(t) = (t,\gamma(t))$ be a C^2-parametrization of the general fold of f such that $\phi(0) = p$ where p is a singular point of f.

(1) p is a fold point of f if $\frac{d}{dt}(f \circ \phi)(0) \neq 0$

(2) p is a cusp point of f if $\frac{d}{dt}(f \circ \phi)(0) = 0$ but $\frac{d^2(f \circ \phi)}{dt^2}(0) \neq 0$.

It is clear that these definitions are independent of the parametrization for the general fold C. Thus we are entitled to adopt the following notations:

$$\frac{df}{dt} = \frac{d(f \circ \phi)}{dt}, \qquad \frac{d^2 f}{dt^2} = \frac{d^2(f \circ \phi)}{dt^2}.$$

Thus p is a fold point of f if $\frac{df}{dt} \neq 0$ at p, p is a cusp point of f if $\frac{df}{dt} = 0$ at p but $\frac{d^2 f}{dt^2} \neq 0$ at p.

Now let us review the two examples again.

In Example 5.1, $f(x,y) = (x^2,y)$, the Jacobian matrix is given by $Jf(x,y) = \begin{pmatrix} 2x & 0 \\ 0 & 1 \end{pmatrix}$, so that $J(x,y) = 2x$. The general fold is y-axis (where $x = 0$).

Let $\phi(t) = (0,t)$, $f \circ \phi(t) = (0,t)$, $\frac{df}{dt}(0,t) = (0,1) \neq 0$, where $(0,t)$ is a singular point of f for every t. Hence, for any t, $(0,t)$ is a fold point of f.

In Example 5.2, $g(x,y) = (xy - y^3,y)$. Then $Jg(x,y) = \begin{pmatrix} y - 3x^2 & x \\ 0 & 1 \end{pmatrix}$ so that $J(x,y) = y - 3x^2$. The general fold C is the parabola $y = 3x^2$. Let C be parametrized by $\phi(t) = (t,3t^2)$. Then $g \circ \phi(t) = (2t^3,3t^2)$ and $\frac{dg}{dt}(t) = (6t^2,6t)$. For any $t \neq 0$, the point $\phi(t)$ is a fold point of g. However for $t = 0$, that is, at $P = (0,0)$, $\frac{dg}{dt} = (0,0)$ and $\frac{d^2 g}{dt^2}(0) = (12t,6)|_{t=0} = (0,6) \neq 0$. By definition then, $(0,0)$ is a cusp point.

Remark 5.2. The parametrization $\phi(t)$ of the general fold C, in either example, does not have any singular point (since $\frac{d\phi}{dt} \neq 0$). However, in the second example, the image of ϕ under g does have a singularity at $(0,0)$.

(There is no such point in Example 5.1 .) The graph for the image of $\emptyset$ under g is as Figure 5.4, where

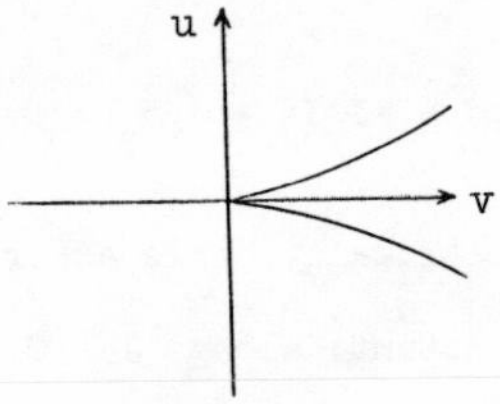

Figure 5.4.

$g(x,y) = (u,v)$. This is why we used the name "cusp" to describe the point (0,0) in the second example, and in the general definition. (cf. section 4 in Chapter 4 as well.)

Remark 5.3. Any singular point of f in Example 5.1 is good since $\nabla J = (2,0) \neq 0$. Similarly, any singular point of g in Example 5.2 is good, since, in this case $\nabla J = (-6x,1) \neq 0$.

Remark 5.4. We shall show that mapping from $\mathbb{R}^2$ to $\mathbb{R}^2$ whose singularities are either folds or cusps form a very large set: any mapping from $\mathbb{R}^2$ to $\mathbb{R}^2$ can be approximated by such a mapping; and such mappings are stable. This is why we restrict our attention to such mappings here.

Next, let us characterize fold and cusp points by means of directional derivatives. From their definition it is clear that

(i) P is a fold point of the mapping $f: \mathbb{R}^2 \to \mathbb{R}^2$ if $f \circ \emptyset$ has non-zero tangent vectors at P .

(ii) P is a cusp point of the mapping $f: \mathbb{R}^2 \to \mathbb{R}^2$ if the tangent vector of $f \circ \emptyset$ at P is zero, but becomes nonzero as we move away from P along the general fold C . It is apparent from this characterization that cusp points are isolated.

Let us now define a vector field W by

$$W(p) = (-J_y(p), J_x(p)) . \tag{5.13}$$

This vector field is tangent to the general fold of f since the directional derivative of the general fold in the W direction is

$$\nabla_W J(p) = -J_y J_x + J_x J_y = 0 \qquad (5.14)$$

along the general fold C. Geometrically, this assertion is also obvious. Consider the level curves $J =$ constant; along $J = 0$, $\nabla J = (J_x, J_y)$ is the normal vector to the level curve $J = 0$, the vector $W(p)$ in (5.13) is just the vector obtained by rotating the vector ∇J through 90°. Thus, W is the tangent vector field of level curves $J =$ constant and in particular of C.

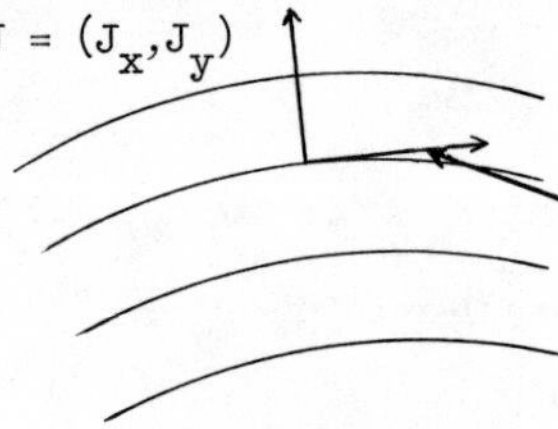

Figure 5.5.

With the previous parametrization ϕ of C, we know that in case P is good,

$$\frac{d\phi}{dt}(t) = W(\phi(t)) \neq \overline{0}. \qquad (5.15)$$

From (5.10) and (5.11) we have:

(1) P is a fold if $\nabla_W f(p) = \frac{df}{dt}(t) \neq \overline{0}$ (5.16)

(2) P is a cusp is $\nabla_W f(p) = \overline{0}$, but $\nabla_W \nabla_W f(p) \neq \overline{0}$ (5.17)

where W is given by (5.13).

With this characterization, let us look at our two examples again.

For Example 5.1,

$$\begin{cases} u = x^2 \\ v = y \end{cases}.$$

$J = 2x$, $J_x = 2$ and $J_y = 0$, thus $W = (0,2)$. Using Equation (5.12), we have

$$\nabla_W f(p) = 0(u_x, v_x) + 2(u_y, v_y) = 2(0,1) = (0,2) \neq (0,0) \tag{5.18}$$

for any good point of f (i.e. for any point (x,y) where $x \neq 0$). Hence any good point is a fold.

For Example 5.2,

$$\begin{cases} u = xy - x^3 \\ v = y \end{cases} .$$

$J = y - 3x^2$, $J_x = -6x$ and $J_y = 1$, thus $W = (-1,-6x)$. Using (5.12), we have

$$\begin{aligned} \nabla_W f(p) &= (-1)(y - 3x^2, 0) - 6x(x,1) \\ &= (-y - 3x^2, -6x) \neq (0,0) , \end{aligned} \tag{5.19}$$

if $(x,y) \neq (0,0)$. However

$$\nabla_W f(0,0) = (0,0) ,$$

but

$$\begin{aligned} \nabla_W \nabla_W f(0,0) &= (-1)(-6x,-6)\big|_{(0,0)} - 6x(-1,0)\big|_{(0,0)} \\ &= (12x,6)\big|_{(0,0)} = (0,6) \neq (0,0) . \end{aligned} \tag{5.20}$$

Hence $(0,0)$ is a cusp point.

<u>Definition 5.3</u>. Let f be a good mapping. A point P in the domain of f is an excellent point if it is regular or a fold point or a cusp point.

The map $f: U \to R^2$, where U is open in R^2, is <u>excellent</u> if each $p \in U$ is excellent.

As we indicated in Remark 5.4 above, we will show that the set of all smooth excellent maps is dense in $C^\infty(U, \mathbb{R}^2)$. For this, we will adopt the same technique we used in section 4 of this chapter to show the density property of Morse functions. We will use the values of the partial derivatives of f to find a bad set in the jet space and prove each of them is small in dimension in comparing with that of the jet space. Then Thom's Transversality Theorem will yield the conclusion that excellent maps are dense. Thus we must next characterize fold and cusp points in terms of partial derivatives.

Let P be a singular point of the good mapping f. Then $\dim \ker(df(p)) = 1$. This means that in the usual cartesian coordinates, the 2×2 matrix of $df(p)$ takes the form

$$A = \begin{pmatrix} a & b \\ c & d \end{pmatrix} \tag{5.21}$$

with (1) $ad - bc = 0$ and (2) not all a, b, c, d zero. Hence their exists non-singular matrices Q and R such that $RAQ = \begin{pmatrix} 0 & 0 \\ 0 & 1 \end{pmatrix}$. Thus, if P is a singular point of a good mapping, we can certainly find coordinates (x,y) and (u,v), in the domain and range spaces respectively, such that

$$Jf(p) = \begin{pmatrix} 0 & 0 \\ 0 & 1 \end{pmatrix}, \tag{5.22}$$

i.e. $u_x = u_y = v_x = 0$, $v_y = 1$. Geometrically speaking this is also quite clear. Since P is a singular point of f, there exists a vector in U mapped into $\overline{0}$ by $\nabla f(p)$. Let the x-axis be in the direction of this vector. The fact that f is good implies that the unit vector in the y-direction is mapped into a nonzero vector, let this vector be in the v-direction. If we normalize it we have (5.22). Using such a coordinate system, we shall find the conditions that P be a fold point or a cusp point.

Using (5.13) and (5.22)

$$\begin{aligned} W = (-J_y, J_x) &= (-(u_x v_y - v_x u_y)_y, (u_x v_y - v_x u_y)_x) \\ &= (-u_{xy}, u_{xx}) \end{aligned} \tag{5.23}$$

at P. Hence by (5.12) and (5.22),

$$\nabla_W f(p) = -u_{xy}(u_x, v_x) + u_{xx}(u_y, v_y) = (0, u_{xx}) \tag{5.24}$$

at P. Therefore, by (5.16), the <u>condition for P to be a fold point</u> is

$$J_x(p) = u_{xx}(p) \neq 0 . \tag{5.25}$$

Now suppose $\nabla_W f(p) = 0$, i.e., $J_x(p) = u_{xx}(p) = 0$. It is clear that

$$J_y = u_{xy} \tag{5.26}$$

at P. And

$$J_{xx} = ((u_x v_y - v_x u_y)_x)_x = u_{xxx} v_y - 2u_{xy} v_{xx} \tag{5.27}$$

at P. Thus, at this point, we have

$$\begin{aligned}
\nabla_W \nabla_W f(p) &= (\nabla_W \nabla_W u,\ \ \nabla_W \nabla_W v)(p) \\
&= (0, -J_y(-J_y v_x + J_x v_y)_x + J_x(-J_y v_x + J_x v_y)_y)(p) \\
&= (0, -u_{xy}(-J_y v_{xx} + J_{xx} v_y))(p) \\
&= (0, -u_{xy}(u_{xxx} - 3u_{xy} v_{xx}))(p) .
\end{aligned} \tag{5.28}$$

Therefore, <u>the condition that P be a cusp point</u> is:

$$\begin{cases} u_{xx} = 0 \\ u_{xy} \neq 0 \\ u_{xxx} - 3u_{xy} v_{xx} \neq 0 \end{cases} \tag{5.29}$$

at the singular point P. It is trivial to check by means of this criterion, that for any y, the point $(0,y)$ is a fold point of the map $u = x^2$, $v = y$ and $(0,0)$ is a cusp point of the map $u = xy - x^3$, $v = y$.

6. <u>Whitney's Theorem</u>

We will state and prove Whitney's Theorem in terms of C^∞-mappings. In doing so we will lose nothing of the spirit of Whitney's Theorem but we will avoid having to fuss about the degree of differentiability of the mappings.

<u>Theorem 6.1</u>. The set of all smooth excellent mappings: $U \to \mathbb{R}^2$, with U open in $\mathbb{R}^2$ is dense in the set $C^\infty(U,\mathbb{R}^2)$.

<u>Theorem 6.2</u>. Let p be fold point of $f \in C^\infty(U,\mathbb{R}^2)$, $U \subset \mathbb{R}^2$. Then smooth coordinate systems (x,y), (u,v) may be introduced about p and $f(p)$ respectively, in terms of which f takes the form:

$$u = x^2, \quad v = y.$$

<u>Theorem 6.3</u>. Let p be a cusp point of $f \in C^\infty(U,\mathbb{R}^2)$, $U \subset \mathbb{R}^2$. Then smooth coordinate systems (x,y), (u,v) may be introduced about p and $f(p)$ respectively, in terms of which f takes the form

$$u = xy - x^3, \quad v = y.$$

We will give Whitney's proof of Theorems 6.1 and 6.2 here. To prove Theorem 6.3, the technique is to use repeatedly the proof of Theorem 6.2 and the following Lemma 6.4. We refer the reader to Whitney's original proof in [87].

<u>Lemma 6.4</u>. Let $g: U \to V$ be a smooth (C^r-) mapping, where U is a neighborhood of $(0,0)$ in $\mathbb{R}^2$ and V is a neighborhood of 0 in $\mathbb{R}$. If $g(0,y) = 0$, $g_x(0,y) = 0$, $g_{xx}(0,y) \neq 0$, then there exists a smooth (C^{r-2}-) mapping $\varphi: U \to V$ such that $\varphi(0,0) \neq 0$ and $g(x,y) = x^2\varphi(x,y)$ in a possible smaller neighborhood of $(0,0) \in U$.

Using Taylor's formula, the proof is straight-forward, and is left to the reader.

7. <u>The Proof of Theorem 6.1</u>

Given any $f_0: U \to \mathbb{R}^2$, where U is open in $\mathbb{R}^2$, we can assume $f_0(\bar{0}) = \bar{0}$ (since we are interested in the local situation), and we will show that there is an excellent map $f: U \to \mathbb{R}^2$ which is arbitrarily close to f_0. According to the characterizations and remarks which have been made in section 5, we know that an excellent mapping $f\colon U \to \mathbb{R}^2$ can be characterized by the values of its partial derivatives of order ≤ 3. Thus, we only have to find the bad set S in $J^3(2,2)$ and show that the codimension of S in $J^3(2,2)$ is greater than the dimension of U, which is two.

For a better understanding, let us illustrate how to find the bad sets S_0, S_1 in $J^1(2,2)$ and $J^2(2,2)$ respectively before we find the bad set S_2 in $J^3(2,2)$.

(1) One of the conditions for f to be excellent is that rank $df \geq 1$ at every point in U, i.e. we cannot have a mapping f sending (x,y) to (u,v) with $u_x = u_y = v_x = v_y = 0$. The bad set S_0 in $J^1(2,2) \cong \mathbb{R}^4$ is precisely the point $\{0,0,0,0\}$ in $J^1(2,2)$. Clearly, codim S_0 in $J^1(2,2) = 4 > 2$.

(2) An excellent mapping is good and hence either $J \neq 0$ or $\nabla J \neq 0$. Thus the bad set S_1 in $J^2(2,2) \cong \mathbb{R}^{10}$ corresponds to the equations $J = 0$ and $\nabla J = 0$, i.e. $J = 0$, $J_x = 0$ and $J_y = 0$. In terms of the α-notations of section 4, these equations may be written:

$$\alpha_1\alpha'_2 - \alpha_2\alpha'_1 = 0. \tag{7.1}$$

$$\alpha_{11}\alpha'_2 + \alpha_1\alpha'_{21} - \alpha_{21}\alpha'_1 - \alpha_2\alpha'_{11} = 0. \tag{7.2}$$

$$\alpha_{12}\alpha'_2 + \alpha_1\alpha'_{22} - \alpha_{22}\alpha'_1 - \alpha_2\alpha'_{12} = 0. \tag{7.3}$$

Let $S_1^{(0)}$ be a subset of S_1 such that $\alpha_1 = \alpha_2 = \alpha'_1 = \alpha'_2 = 0$ (i.e. $u_x = u_y = v_x = v_y = 0$). Codim $S_1^{(0)} > 2$, which follows from the same argument as in the preceding paragraph. Let $S_1^{(1)} = S_1 - S_1^{(0)}$. We claim codim $S_1^{(1)} > 2$. In order to prove this, we differentiate (7.1), (7.2) and (7.3) with respect to $\alpha_1,\alpha_2,\ldots,\alpha_{11},\ldots,\alpha'_{22}$. We obtain the gradient vectors:

$$\nabla J(\alpha) = (\alpha'_2,-\alpha'_1,0,0,0,-\alpha_2,\alpha_1,0,0,0) \tag{7.4}$$

$$\nabla(J_x)(\alpha) = (\alpha'_{21},-\alpha'_{11},\alpha'_2,-\alpha'_1,0,-\alpha_{21},\alpha_{11},\alpha_2,-\alpha_1,0) \tag{7.5}$$

$$\nabla(J_y)(\alpha) = (\alpha'_{22},-\alpha'_{12},0,\alpha'_2,-\alpha'_1,-\alpha_{22},\alpha_{12},0,-\alpha_2,\alpha_1). \tag{7.6}$$

For any point in $S_1^{(1)}$, at least one of $\alpha_1, \alpha_2, \alpha_1', \alpha_2'$ is nonzero, say $\alpha_2' \neq 0$. α_2' appears in (7.4) at the first component, appears in (7.5) at the third component and appears in (7.6) at the fourth component. Thus, these three gradient vectors are linearly independent. This last statement is also true for either $\alpha_1 \neq 0$, $\alpha_2 \neq 0$ or $\alpha_1' \neq 0$ by the same reason as above. Hence $S_1^{(1)}$ is a 7-dimensional manifold which is of course of codimension three in $J^2(2,2)$.

(3) A similar argument will be used to prove that the bad set S_2 in $J^3(2,2) \cong \mathbb{R}^{18}$ is also of codimension > 2. According to the characterization of folds and cusps in section 5 of this chapter, the bad set S_2 could be written as $S_2^{(0)} \cup S_2^{(1)} \cup S_2^{(2)}$ where

$$S_2^{(0)} = \{(\alpha_1,\alpha_2,\ldots,\alpha_{111},\alpha_{112},\ldots,\alpha_{222}') \mid \alpha_1 = \alpha_2 = \alpha_1' = \alpha_2' = 0\} \tag{7.7}$$

$$S_2^{(1)} = \{\alpha = (\alpha_1,\ldots,\alpha_{222}') \mid \alpha \notin S_2^{(0)} \text{ and } \alpha \text{ satisfies (7.1), (7.2) and (7.3)}\} \tag{7.8}$$

$$S_2^{(2)} = \{\alpha \mid \alpha \notin S_2^{(0)} \cup S_2^{(1)} \text{ and } \alpha \text{ satisfies } J = 0, \nabla_W f = 0 \text{ and } \nabla_W \nabla_W f = 0\} \tag{7.9}$$

It is clear from (1), (2) that the codimensions of $S_2^{(0)}$ and $S_2^{(1)}$ are greater than two. We will show that $\text{cod } S_2^{(2)} = 3 > 2$.

For any point $\alpha \in S_2^{(2)}$, one of $\alpha_1, \alpha_2, \alpha_1'$ and α_2' is not zero and furthermore, either $J_x \neq 0$ or $J_y \neq 0$. Without loss of generality, let $\alpha_2' \neq 0$. In this case it is obvious, from the explanation of the last section, that the equations $\nabla_W f = 0$ and $\nabla_W \nabla_W f = 0$ are the same as $\nabla_W v = 0$ and $\nabla_W \nabla_W v = 0$ (since the equations $\nabla_W u = 0$, $\nabla_W \nabla_W u = 0$ are consequences of the last two equations for ∇f of rank 1 when $J = 0$, and $\alpha_2' \neq 0$ corresponds to $v_y \neq 0$). Thus, to show $\text{codim } S_2^{(2)} = 3$ it is equivalent to show that $J = 0$, $\nabla_W v = 0$ and $\nabla_W \nabla_W v = 0$ have linearly independent gradients. Let

$$F_1(\alpha) = \alpha_1\alpha_2' - \alpha_2\alpha_1' \tag{7.10}$$

correspond to J. Let

$$F_2(\alpha) = -(\alpha_{12}\alpha_2' + \dots - \alpha_2\alpha_{12}')\alpha_1' + (\alpha_{11}\alpha_2' + \dots - \alpha_2\alpha_{11}')\alpha_2' \\ = \alpha_{11}(\alpha_2')^2 + \dots \tag{7.11}$$

correspond to the gradient $\nabla_W v$ which is $-J_y v_x + J_x v_y$. In (7.11), (7.2) and (7.3) has been used. Let

$$F_3(\alpha) = 0$$

correspond to the equation

$$\nabla_W \nabla_W v = -J_y(-J_y v_x + J_x v_y)_x + J_x(-J_y v_x + J_x v_y)_y = 0 . \tag{7.12}$$

By studying (7.12) in detail in terms of the α's, the terms involving α_{111} and α_{112} are most interesting to us. Rearranging the terms, we have

$$\begin{aligned} F_3(\alpha) = & -(\alpha_{12}\alpha_2' + \dots - \alpha_2\alpha_{12}')(\alpha_2')^2\alpha_{111} \\ & + [2(\alpha_{12}\alpha_2' + \dots - \alpha_2\alpha_{12}')\alpha_1'\alpha_2' \\ & + (\alpha_{11}\alpha_2' + \dots - \alpha_2\alpha_{11}')(\alpha_2')^2]\alpha_{112} + \dots = 0 . \end{aligned} \tag{7.13}$$

Now, let us observe the following equations:

$$\nabla F_1(\alpha) = (\alpha_2', -\alpha_1', 0, 0, \dots; -\alpha_2, \alpha_1, 0, 0, \dots) \tag{7.14}$$

$$\frac{\partial F_2}{\partial \alpha_{11}}(\alpha) = (\alpha_2')^2 \neq 0 \tag{7.15}$$

$$\frac{\partial F_3}{\partial \alpha_{112}}(\alpha) = 2(\alpha_{12}\alpha_2' + \dots - \alpha_2\alpha_{12}')\alpha_1'\alpha_2' + \\ + (\alpha_{11}\alpha_2' + \dots - \alpha_2\alpha_{11}')(\alpha_2')^2 . \tag{7.17}$$

∇F_1, ∇F_2 are nonzero since $\alpha_2' \neq 0$, furthermore, since one of $(\alpha_{12}\alpha_2' + \dots - \alpha_2\alpha_{12}')$, $(\alpha_{11}\alpha_2' + \dots - \alpha_2\alpha_{11}')$ is nonzero, then $\nabla F_3 \neq 0$. In fact ∇F_1, ∇F_2 and ∇F_3 are linearly independent since α_2' appears in different columns of the three gradient vectors ∇F_1, ∇F_2 and ∇F_3. Therefore cod $S_2^{(2)} = 3$. For the same reason as in section 4 of this chapter S_0, $S_0 \cup S_1$,

$S_0 \cup S_1 \cup S_2$ are closed sets which means that $\{S_0, S_1, S_2\}$ is a stratified manifold. Thus, in view of Theorem 3.1, Theorem 6.1 is proved.

8. The Proof of Theorem 6.2

Suppose f is smooth and good, suppose $(0,0) = \overline{0}$ is a fold point of f. Our purpose in this section is to introduce coordinates in both domain and range of f at $\overline{0} \in U$ and $\overline{0} \in \mathbb{R}^2$ such that f takes the form

$$\begin{aligned} u &= x^2 \\ v &= y \;. \end{aligned} \tag{8.1}$$

First of all, let us expand u, v around $\overline{0}$:

$$\begin{aligned} u(x,y) &= u_x(\overline{0})x + u_y(\overline{0})y + a_1x^2 + a_2xy + a_3y^2 + \cdots \\ v(x,y) &= v_x(\overline{0})x + v_y(\overline{0})y + b_1x^2 + b_2xy + b_3y^2 + \cdots \end{aligned} \tag{8.2}$$

where $a_1 = \frac{1}{2}u_{xx}(\overline{0})$, $a_2 = u_{xy}(\overline{0})$, $a_3 = \frac{1}{2}u_{yy}(\overline{0})$, $b_1 = \frac{1}{2}v_{xx}(\overline{0})$, $b_2 = v_{xy}(\overline{0})$, $b_3 = \frac{1}{2}v_{yy}(\overline{0})$, etc. Referring to the coordinates (x,y) and (u,v) used in (5.22), we have $u_x = u_y = v_x = 0$ and $v_y = 1$ at $\overline{0}$, hence (8.2) can be simplified to:

$$\begin{aligned} u(x,y) &= a_1x^2 + a_2xy + a_3y^2 + \cdots \\ v(x,y) &= y + b_1x^2 + b_2xy + b_3y^2 + \cdots \;. \end{aligned} \tag{8.3}$$

Let us first lay out the plan of the rest of the argument and then the details will be filled in. We are trying to find coordinates (x',y') in a neighborhood of $\overline{0} \in U$ such that $v_{x'} = 0$, $v_{y'} = 1$ and then change coordinates in a neighborhood of $\overline{0}$ of the range space into u' and $v' = v$ such that $u'_{x'} = 0$, $u'_{x'x'} \neq 0$ at $\overline{0}$. If we can do so, we have

$$J(\overline{0}) = u'_{x'}v'_{y'} - v'_{x'}u'_{y'} = 0 \tag{8.4}$$

and

$$J_x(\bar{0}) = u'_{x'x'}(\bar{0}) \neq 0 . \tag{8.5}$$

Equations (8.3) in this new coordinate system give us that

$$u'(x',y') = (x')^2 \eta(x',y') \tag{8.6}$$

since a_2, a_3 and many others are zero. Here the crucial point is that $\eta(0,0) \neq 0$ by Lemma 6.4 . Granting this, Theorem 6.2 is proved by the following coordinate change. Let

$$\begin{aligned} x^* &= x' \, \eta(x',y')^{1/2} \\ y^* &= y' . \end{aligned} \tag{8.7}$$

Then, we have

$$\begin{aligned} u' &= (x^*)^2 \\ v' &= v = y' = y^* \end{aligned} \tag{8.8}$$

which is of the form (8.1) .

For the details of finding (x,y), and (u',v'), we let

$$\begin{aligned} \bar{x} &= x \\ \bar{y} &= y + b_1 x^2 + b_2 xy + b_3 y^2 + \ldots \end{aligned} \quad . \tag{8.9}$$

This is a valid coordinate change since

$$\begin{vmatrix} \frac{\partial \bar{x}}{\partial x} & \frac{\partial \bar{x}}{\partial y} \\ \frac{\partial \bar{y}}{\partial x} & \frac{\partial \bar{y}}{\partial y} \end{vmatrix} (\bar{0}) = 1 \neq 0 .$$

Now the problem is to write u in these new coordinates. That is, to invert the coordinates between (x,y) and $(\bar{x},\bar{y})$ for the function u . Consider

$$G(x,y,\bar{y}) = \bar{y} - y - b_1 x^2 - b_2 xy - b_3 y^2 - \ldots \tag{8.10}$$

as a function of three variables. Then we are interested in the surface $G = 0$ and on the surface (i.e. when $\bar{y} = y + b_1x^2 + b_2xy + b_3y^2 + \ldots$) we shall solve for y in terms of x and $\bar{y}$. By the observation $G_y(0,0,0) = -1 \neq 0$, we can use the Implicit Function Theorem to obtain the existence of a map $y = h(x,\bar{y})$ such that $G(x,h(x,\bar{y}),\bar{y}) = 0$. Hence

$$y = \bar{y} - b_1x^2 - b_2xh(x,\bar{y}) - b_3(h(x,\bar{y}))^2 - \ldots \tag{8.11}$$

on this surface. Since $x = \bar{x}$, we can get our inverse transformation

$$(\bar{x},\bar{y}) \to (x,y) = (\bar{x},\bar{y} - \bar{S}(\bar{x},\bar{y})) \tag{8.12}$$

where $\bar{S}(\bar{x},\bar{y}) = b_1\bar{x}^2 + b_2\bar{x}h(\bar{x},\bar{y}) + b_3(h(\bar{x},\bar{y}))^2 \ldots$ Thus,

$$u = u(x(\bar{x},\bar{y}),y(\bar{x},\bar{y})) = u(\bar{x},\bar{y} - \bar{S}(\bar{x},\bar{y})\,. \tag{8.13}$$

We now define

$$\bar{u}(\bar{x},\bar{y}) = u(\bar{x},\bar{y} - \bar{S}(\bar{x},\bar{y}))$$

and (8.14)

$$\bar{v}(\bar{x},\bar{y}) = v(x(\bar{x},\bar{y}),y(\bar{x},\bar{y}))\,.$$

Then

$$\begin{aligned}\bar{v}(\bar{x},\bar{y}) &= y(\bar{x},\bar{y}) + b_1(x(\bar{x},\bar{y}))^2 + b_2x(\bar{x},\bar{y})y(\bar{x},\bar{y}) + b_3(y(\bar{x},\bar{y}))^2 + \ldots \\ &= \bar{y} - b_1x^2 - b_2xh(x,\bar{y}) - b_3(h(x,\bar{y}))^2 \ldots \\ &\quad + b_1x^2 + b_2xy(\bar{x},\bar{y}) + b_3(y(\bar{x},\bar{y}))^2 + \ldots = \bar{y}\,,\end{aligned} \tag{8.15}$$

since $y = h(x,\bar{y}) = h(\bar{x},\bar{y})$.

With this new coordinate system $\bar{v}_{\bar{x}} = 0$, $\bar{v}_{\bar{y}} = 1$, then $J(\bar{x},\bar{y}) = \bar{U}_{\bar{x}}(\bar{x},\bar{y})$. The curve of fold points is given by $J = 0$, or in our case $\bar{U}_{\bar{x}}(\bar{x},\bar{y}) = 0$. Since $\bar{0}$ is a fold point, $J_x(\bar{0}) = \bar{U}_{xx}(\bar{0}) \neq 0$. By the Implicit Function Theorem, we may solve $J = 0$ near $\bar{0}$ and a smooth function Ψ can be obtained such that

$$J(\Psi(\overline{y}),\overline{y}) = 0 \tag{8.16}$$

for $\overline{y}$ near 0 in $\mathbb{R}$.

Notice that if we are eventually going to get coordinates in which f has the normal form, the fold must be mapped to the second coordinate axis. Hence, our next transformation is designed to ensure just that. Set

$$\begin{aligned} x' &= \overline{x} + \Psi(\overline{y}) \\ y' &= \overline{y} \end{aligned} \tag{8.17}$$

and

$$\begin{aligned} u'(x',y') &= \overline{u}(\overline{x} + \Psi(\overline{y}),\overline{y}) - \overline{u}(\Psi(\overline{y}),\overline{y}) \\ v'(x',y') &= \overline{y} = y' , \end{aligned} \tag{8.18}$$

then

$$u'(0,y) = 0 \tag{8.19}$$

and also

$$u'_{x'}(0,y') = \overline{u}_{\overline{x}}(\Psi(\overline{y}),\overline{y}) = 0 . \tag{8.20}$$

Further, since

$$\overline{u}_{\overline{xx}}(0) = J_{\overline{x}}(\overline{0}) \neq 0 ,$$

$$u'_{x'x'}(\overline{0}) = \overline{u}_{xx}(\Psi(0),0) = \overline{u}_{xx}(0,0) \neq 0 .$$

By Lemma 6.4

$$u'(x',y') = (x')^2 \eta(x',y')$$

with $\eta(0,0) \neq 0$, as required.

CHAPTER 3

UNFOLDINGS OF MAPPINGS

1. Introduction

In the opening paragraph of Chapter 1, we mentioned three basic ideas in singularity theory. So far we have discussed (1) stability and (2) genericity. In this chapter, we will concentrate on the third basic idea in this theory, namely the unfolding of singularities. We would like to indicate why it is one of the basic ideas in singularity theory before we go into a detailed discussion of precisely what the unfolding is.

After reading the first two chapters, we believe that the reader will realize the importance of the concept of stability. It is worthwhile to reemphasize its importance here. In any branch of science, it is always a challenge to try to classify the objects under study. Unfortunately, it is often extremely difficult to carry out this classification. It becomes much easier if one tries to classify only the stable objects. It is also important to point out that in many cases the stable objects are generic, thus, in these cases, every object is either stable or close to a stable one. Another reason for the importance of stability is that, due to the introduction of the theory of catastrophes, the theory of singularities has acquired many important applications to the natural sciences. This is specially true in the area of biology - particularly in developmental biology - linguistics, economics, fluid and gas dynamics, the buckling problems in engineering, the study of heartbeat and of nerve impulse, as well as in the sociological and psychological sciences. Stability is a natural condition to place upon mathematical models for processes in nature because the conditions under which such processes take place can never be exactly duplicated; therefore what is observed must be invariant under small perturbations and hence stable.

On the other hand, stable objects have boundaries where discontinuities appear. We all know that mathematics used in almost all sciences so far is based on the differential calculus, which presupposes continuity. There is a great demand, therefore for a mathematical theory to explain and predict (if possible) the occurrence of discontinuous phenomena. Thom's theory of catastrophes endeavors to provide a framework for the discussion of those situations, among the most important in nature, in which a continuous change or perturbation of the control variables leads to a discontinuous change in outcome. Most importantly, in the early 1960's, Thom realized that elementary catastrophes, which are certain singularities of smooth maps $\mathbb{R}^r \to \mathbb{R}^r$ with $r \leq 4$, could be finitely classified by unfolding certain polynomial germs $(x^3, x^4, x^5, x^6, x^3 + y^3, x^3 - xy^2$ and $x^2y + y^4)$.

The models generated by elementary catastrophes are arousing considerable excitement among many scientists today. Let us consider a biological (or chemical) system in space-time $\mathbb{R}^4$, where each spatial locus is presumed to be a cell developing over the time t. Let the biochemical states of the cell lie in a euclidean space $\mathbb{R}^n$. Let us assume that we are given a potential function $F: \mathbb{R}^n \times \mathbb{R}^4 \to \mathbb{R}$ which calculates the local thermodynamical advantages for a locus $\overline{u} \in \mathbb{R}^3$ at time t to be in a state $\overline{x} \in \mathbb{R}^n$. Let us further assume that the system is a minimizing system in the sense that at each point $(\overline{u},t)$, the system will be in those states $\overline{x}$ at which $F(\overline{x},\overline{u},t)$ is minimized. When we ask ourselves what kind of discontinuities may appear, we are asking the equivalent mathematical question of what kind of discontinuities the projection map

$$\pi: \{\overline{x} = (x_1,\ldots,x_n) \in \mathbb{R}^n: \frac{\partial F(\overline{x})}{\partial x_i} = 0 \quad 1 \leq i \leq n\} \times \mathbb{R}^4 \to \mathbb{R}^4 \tag{1.1}$$

can possess, for they will occur at the positions in space-time at which the system cannot choose in a smooth manner the state $\overline{x}$ in which it is to be. Thus, we are required to analyze the possible discontinuities of such a projection map. To do so, we must translate the physical nuances of the problem to mathematical ones: F is, first of all, a smooth map considered as a map $F: \mathbb{R}^n \times \mathbb{R}^4 \to \mathbb{R}$. Denote the origin of $\mathbb{R}^4$ by $\overline{0}$. At $\overline{0}$, $F_{\overline{0}}: \mathbb{R}^n \to \mathbb{R}$, $F_{\overline{0}}(\overline{x}) = F(\overline{x},\overline{0})$, describes the thermodynamical advantage of the system initially and as the process continues the function $F_{(\overline{u},t)}: \mathbb{R}^n \to \mathbb{R}$ changes gradually.

Note that the question of that we observe is independent of what coordinate system (i.e., method of measurement) we use on $\mathbb{R}^n$ as long as we go from one set of coordinates to the other set smoothly. Similarly, how we coordinatize $\mathbb{R}^4$ locally is irrelevant as long as we do so smoothly. This is because we are dealing with questions of existence of minima points in $\mathbb{R}^n$ and of singularities of π.

It is plain that the question we have asked is too general and as such we probably will not be able to answer it very easily. But then the remarkable question (almost fantasy) by René Thom arises: can we find a finite number of relatively simple functions F for which the study of the projection π is tractable and to which the study of many other functions (a dense set) is related by means of a simple operation?

As we shall see, the answer is yes. How René Thom went about answering it, really is the mathematical essence of his "Catastrophe Theory" and is what we will discuss in this chapter. If we extract from the rest of this chapter, the heart of the proof of Thom's Classification Theorem lies in the concept of the stability of unfoldings, which is a key mathematical concept in catastrophe theory. However it is important to emphasize at this point that the theory of stable unfoldings is distinct from the theory of stable germs. With regard to the stability of unfoldings, those conjugating diffeomorphisms will be required to respect the fibration of $\mathbb{R}^n \times \mathbb{R}^r$ by fibres $\mathbb{R}^n \times \overline{u}$, $\overline{x} \times \mathbb{R}^r$. Thus, we will introduce the basic concept of the stability of the universal unfolding of a singularity in this chapter. We will further indicate some of the important results associated with this concept, leading to a description of the seven elementary catastrophes. We will require several theorems with quite lengthy proofs. We will give appropriate references instead of giving the details of the proofs. However, what we will emphasize in this chapter is the intuitive, geometrical description of the definitions involved in those theorems. The examples should be helpful to the reader in obtaining a better understanding of those definitions.

2. Germs of Mappings

Our main interest in this chapter will be in the local properties of maps. We begin this section by recalling some elementary definitions of a local nature.

Definition 2.1. Let S be the set of all continuous (C^0-) maps: $\mathbb{R}^n \to \mathbb{R}^p$ defined in a neighborhood of the origin. We say that two such C^0- maps, $f,g \in S$ determine the same map-germ (or simply germ) if they agree in some neighborhood of the origin, so that a germ of a C^0- map is, strictly speaking, an equivalence class of C^0- maps. Since our theory is entirely local, we will permit ourselves to speak of the values of a germ $\tilde{f}$ and to write $\tilde{f}(x)$, $x \in \mathbb{R}^n$, although it would be more correct to choose a representative f from the equivalence class $\tilde{f}$. We may also talk of germs: $\mathbb{R}^n \to \mathbb{R}^p$ at points of $\mathbb{R}^n$ different from origin. Notice that here we write $\tilde{f} = [f]$, the equivalence class of f, and sometimes we even do not distinguish the germ $\tilde{f}$ and its representatives.

A germ $\tilde{f}$ at x is smooth or C^∞ (analytic or C^ω) if it has a representative which is smooth (analytic respectively) in a neighborhood of x.

Germs behave much the same as maps. For example they can be composed the same way as maps, i.e. if $\tilde{f}: \mathbb{R}^n \to \mathbb{R}^p$ is a germ at $x \in \mathbb{R}^n$ and $\tilde{g}: \mathbb{R}^p \to \mathbb{R}^m$ is a germ at $\tilde{f}(x) \in \mathbb{R}^p$, the germ of $\tilde{g} \circ \tilde{f}$ at x can be defined in a natural way by taking the equivalence class of the composition of representatives of $\tilde{f}$ and $\tilde{g}$.

Definition 2.2. We denote by $\varepsilon(n,p)$ the set of germs at $\bar{0} \in \mathbb{R}^n$ of smooth maps $\mathbb{R}^n \to \mathbb{R}^p$. If $p = 1$, we shall write $\varepsilon(n)$ for $\varepsilon(n,1)$.

Definition 2.3. $m(n) = \{\tilde{f} \in \varepsilon(n) \mid \tilde{f}(\bar{0}) = 0\}$.

It is clear that $\varepsilon(n)$ is a ring with identity, where the identity is the germ at $\bar{0}$ of the constant function taking the value $1 \in \mathbb{R}$. Addition and multiplication in $\varepsilon(n)$ are induced by the $\mathbb{R}$-algebra structure of $\mathbb{R}$. It is also clear that $m(n)$ is an ideal of $\varepsilon(n)$. Given $\tilde{f}_1,\ldots,\tilde{f}_r \in \varepsilon(n)$, we adopt the following notation: $\langle\tilde{f}_1,\ldots,\tilde{f}_r\rangle_{\varepsilon(n)}$ is the ideal in $\varepsilon(n)$ generated by $\tilde{f}_1,\tilde{f}_2,\ldots,\tilde{f}_r$, i.e. the set of germs is expressible in the form $\sum_{i=1}^{r} \alpha_i\tilde{f}_i$, where

$\alpha_i \in \varepsilon(n)$. More generally, let $\langle e_1,\ldots,e_n\rangle_A$ be the module generated by $\{e_i\}_{i=1,\ldots,n}$, over the ring A.

Let us recall that a local ring is a commutative ring with an identity and with a unique maximal ideal.

Theorem 2.1. $\varepsilon(n)$ is a local ring with the unique maximal ideal $m(n)$, which is generated by the coordinate functions.

Proof: Let us prove the second statement first. Let $x_1,\ldots,x_n$ be the coordinates of $\mathbb{R}^n$ and let $\tilde{f}$ be an element of $m(n)$ with a representative f. Then

$$
\begin{aligned}
f(x) &= \int_0^1 \frac{d}{dt}(f(xt))dt \\
&= \int_0^1 \Big(\sum_{i=1}^{n} x_i \frac{\partial}{\partial y_i} f(y)_{xt}\Big)dt \qquad (2.1)\\
&= \Sigma\, f_i(x)x_i ,
\end{aligned}
$$

where $f_i(x) = \int_0^1 (\frac{\partial f}{\partial y_i}(y)_{xt})dt$. This implies that $\tilde{f} \in \langle \tilde{x}_1,\ldots,\tilde{x}_n\rangle_{\varepsilon(n)}$ or $m(n) \subset \langle \tilde{x}_1,\ldots,\tilde{x}_n\rangle_{\varepsilon(n)}$. The converse inclusion $\langle \tilde{x}_1,\ldots,\tilde{x}_n\rangle_{\varepsilon(n)} \subset m(n)$ is trivial since each $\tilde{x}_i \in m(n)$. To avoid being pedantic, let us identify $\tilde{x}_i$ and x_i, so that we may say that $m(n)$ is generated by the coordinate functions.

Now if $\tilde{f} \in \varepsilon(n)$ and $\tilde{f} \notin m(n)$, then by definition of $m(n)$, $f(\bar{0}) \neq 0$ and we can choose a neighborhood U of $\bar{0} \in \mathbb{R}^n$ such that $f \neq 0$ in U. (Throughout the rest of this book, unless otherwise stated, f shall be considered as a representative of $\tilde{f}$, g be a representative of $\tilde{g}$, etc.) Then $1/f$ exists in U and is C^∞. Let $\tilde{g}$ be the germ $[1/f]$, then $\tilde{f}\cdot\tilde{g} = [f][1/f] = [1] =$ identity in $\varepsilon(n)$. Thus the ideal generated by $m(n)$ and $\tilde{f}$, $\langle m(n),\tilde{f}\rangle_{\varepsilon(n)}$, is $\varepsilon(n)$, so that $m(n)$ is maximal. Finally we want to show it is the unique maximal ideal of $\varepsilon(n)$. Let I be any proper ideal of $\varepsilon(n)$, we claim that $I \subset m(n)$. If $I \not\subset m(n)$, then there is a germ $\tilde{f} \in I - m(n)$. By the same argument we just used, $\tilde{f}$ has an inverse $\tilde{g}$ in $\varepsilon(n)$, and $\tilde{g}\tilde{f} = \mathrm{id} \in I$. Thus $I = \varepsilon(n)$, contrary to the hypothesis that I is a proper ideal of $\varepsilon(n)$.

Now we denote by $m(n)^k$ the set of those germs which vanish at $\bar{0}$ together with their derivatives of order $< k$. In other words, $m(n)^k = \{\tilde{f} \in m(n) \mid j^{k-1}(f) = 0, \forall \; f \in \tilde{f}\}$. It is easy to see that $m(n)^k$ is the ideal generated by all monomials in x_i of degree k. Thus $m(n)^k$ is a finitely generated $\varepsilon(n)$-module. Clearly if $\tilde{f} \in m(n)^2$ then $\tilde{f}$ is singular (at $\bar{0}$); if $\tilde{f} \in m(n)^3$ then $\tilde{f}$ is said to be <u>degenerate</u>.

<u>Definition 2.4</u>. We write $L(n) = \{\varphi \in \varepsilon(n,n) \mid \varphi(\bar{0}) = \bar{0}$ and φ is non-singular at $\bar{0}\}$ for the <u>group</u> of germs of local diffeomorphisms of $\mathbb{R}^n$ at $\bar{0}$ under the operation of composition.

<u>Definition 2.5</u>. Let $\tilde{f}$ and $\tilde{g}$ be germs of $\varepsilon(n)$. We say $\tilde{f}$ and $\tilde{g}$ are <u>right-equivalent</u>, and write $\tilde{f} \sim_r \tilde{g}$ if there is a $\varphi \in L(n)$ such that $\tilde{f} = \tilde{g} \circ \varphi$. We say $\tilde{f}$ and $\tilde{g}$ are <u>right-left equivalent</u> (or <u>2-sided equivalent</u>), $\tilde{f} \sim_{r\ell} \tilde{g}$, if there exist $\varphi \in L(n)$, $\psi \in L(1)$ such that $\tilde{f} = \psi \tilde{g} \varphi$.

It is clear that both right equivalence and right-left equivalence are equivalence relations.

<u>Definition 2.6</u>. Let $\tilde{f} \in m(n)$ and let k be a non-negative integer. Then $\tilde{f}$ is <u>right (or right-left) - determined by its k-jet</u>, or simply <u>right (or right-left) k-determined</u> if, for every $\tilde{g} \in m(n)$ such that $j^k(\tilde{f}) = j^k(\tilde{g})$, then $\tilde{f} \sim_r \tilde{g}$ (respectively $\tilde{f} \sim_{r\ell} \tilde{g}$).

It is clear that the property of being right (or right-left) k-determined is invariant under right (respectively right-left) equivalence. Obviously we have that any right k-determined germ is right-left determined, but the converse is false. Example 2.4 will provide us with a counterexample. With the help of the Implicit Function Theorem, we have one more simple observation, namely any non-singular germ $\tilde{f} \in m(n)$ (i.e. $\tilde{f} \notin m(n)^2$) is right 1-determined, thus all non-singular germs are r-equivalent to each other.

For the rest of this book we will drop the word "right" or "r." Whenever we write k-determined or equivalent $(\sim)$, we mean right k-determined or right equivalent $(\sim_r)$ respectively.

<u>Definition 2.7</u>. A germ $\tilde{f} \in m(n)$ is <u>finitely determined</u> if there is a non-negative integer k such that $\tilde{f}$ is k-determined.

An equivalent way of expressing the fact that $\tilde{f}$ is right-left k-determined is to say that $j^k(f)$ is <u>C^∞-sufficient</u>, any two realizations f_1, f_2 of the k-jet $j^k(f)$ are C^∞-equivalent. Explicitly, there exist local C^∞-diffeomorphisms $h_1 \colon \mathbb{R}^n, \bar{0} \to \mathbb{R}^n, \bar{0}$, and $h_2 \colon \mathbb{R}^1, 0 \to \mathbb{R}^1, 0$, such that the diagram

$$\begin{array}{ccc} (\mathbb{R}^n, 0) & \xrightarrow{\;f_1\;} & (\mathbb{R}^1, 0) \\ {\scriptstyle h_1}\downarrow & & \downarrow{\scriptstyle h_2} \\ (\mathbb{R}^n, 0) & \xrightarrow[\;f_2\;]{} & (\mathbb{R}^1, 0) \end{array}$$

commutes. We may similarly speak of $j^k(f)$ being <u>C^m-sufficient</u> $(m \geq 0)$; the modification required in the definition is evident. Finally we observe that if $j^k(f)$ is not C^m-sufficient, then $\tilde{f}$ is not right-left k-determined and hence not k-determined. The following definition is given for future reference.

<u>Definition 2.8</u>. Let $\tilde{f} \in m(n)$. <u>The degree of C^∞-sufficiency</u> of $\tilde{f}$ is the smallest integer k such that $\tilde{f}$ is right-left k-determined. [37]

<u>Example 2.1</u>. The germ of $f(x,y) = x^2$ is not finitely determined. In other words, for any positive integer k, the k-jet of f is not a C^∞-sufficient k-jet. For there is an integer N such that $2N > k$, and the germ of $g(x,y) = x^2 - y^{2N}$ is not equivalent to $\tilde{f}$ although $j^k(\tilde{f}) = j^k(\tilde{g})$. This example will be used in Example 1.1 of Chapter 6 where we show that f and g are not even C^0-equivalent.

<u>Example 2.2</u>. The germ of $f(x,y) = x^2 + y^3$ is 3-determined. In other words, for any C^∞-function P with $j^3(P) = 0$, $f \sim f + P$. The simple proof of this statement will lead us to the beginning of the proof of the criterion of the

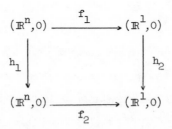

finite-determinancy for germs. For the proof for $f \sim f + P$, we write P as a formal Taylor series expansion:

$$P(x,y) = a_1x^4 + a_2x^3y + a_3x^2y^2 + a_4xy^3 + a_5y^4 + \dots ,$$

$$(f + P)(x,y) = x^2(1 + a_1x^2 + a_2xy + a_3y^2 + \dots) + y^3(1 + a_4x + a_5y + \dots) .$$

Let

$$X(x,y) = x(1 + a_1x^2 + a_2xy + a_3y^2 + \dots)^{1/2} \tag{2.2}$$

$$Y(x,y) = y(1 + a_4x + a_5y + \dots)^{1/3} \tag{2.3}$$

and

$$h(x,y) = (X(x,y),Y(x,y)) . \tag{2.4}$$

Then h is a local C^∞-diffeomorphism $\mathbb{R}^2 \to \mathbb{R}^2$, with $h(\overline{0}) = \overline{0}$, and

$$f(h(x,y)) = f(X(x,y),Y(x,y)) = (f + P)(x,y) \tag{2.5}$$

for all (x,y) in a neighborhood of $\overline{0} \in \mathbb{R}^2$. Thus $f \sim f + P$.

<u>Example 2.3</u>. The germ of $f(x,y) = x^5 + y^5$ is not 5-determined. It is proved in [27] that the 5-jet of f is C^1-sufficient but not C^2-sufficient. Hence $j^5(f)$ is not C^∞-sufficient. Then $\tilde{f}$ is not 5-determined.

<u>Example 2.4</u>. We consider the germ of $f(x,y) = x^2y + H_4(x,y)$, where H_4 is a homogeneous polynomial of degree 4 in two variables x and y. It is proved in [37] that $j^4(f) \sim x^2y + y^4$ (which is Thom's parabolic umbilic) if $H_4(0,1) > 0$ and $j^4(f) \sim -(x^2y + y^4)$ if $H_4(0,1) < 0$. This allows us to conclude that $j^4(f) \sim_{r\ell} x^2y + y^4$ in case $H_4(0,1) \neq 0$, but, in fact, $\tilde{f}$ is not 4-determined in general. In particular, $x^2y + y^4$ is not equivalent to $x^2y - y^4$ although $x^2y + y^4 \sim_{r\ell} x^2y - y^4$.

In order to detect which germs are finitely determined, it is important to have a criterion, and this will be given in the next section.

3. Finitely Determined Germs

For convenience we shall agree that throughout this chapter, unless otherwise stated, x shall denote an element $(x_1,\ldots,x_n) \in \mathbb{R}^n$ with standard coordinates. If $\tilde{f} \in m(n)^2$, then its partial derivatives $\frac{\partial\tilde{f}}{\partial f_i}$, $i = 1,\ldots,n$, belong to $m(n)$ and we write $\underline{\langle\frac{\partial\tilde{f}}{\partial x}\rangle}$ for the ideal generated by the $\frac{\partial\tilde{f}}{\partial x_i}$, or

$$\langle\frac{\partial\tilde{f}}{\partial x}\rangle = \langle\frac{\partial f}{\partial x_1},\ldots,\frac{\partial f}{\partial x_n}\rangle_{\varepsilon(n)} \tag{3.1}$$

It is clear that $\langle\frac{\partial\tilde{f}}{\partial x}\rangle \subseteq m(n)$ if $\tilde{f} \in m(n)^2$.

Example 3.1. Let $f(x_1,x_2) = x_1^2 + x_2^3$, then

$$\langle\frac{\partial\tilde{f}}{\partial x}\rangle = \langle x_1, x_2^2\rangle_{\varepsilon(n)}\,.$$

We now state and prove one of the key theorems in this chapter.

Theorem 3.1. (Mather) Let $\tilde{f} \in m(n)$. Let k be a non-negative integer. If $m(n)^{k+1} \subset m^2\langle\frac{\partial\tilde{f}}{\partial x}\rangle$, then $\tilde{f}$ is k-determined.

Since this is one of the key results in the theory, it is helpful for the reader to know how this criterion has been obtained. The essential idea in proving this criterion is to find a "flow" (C^∞-diffeomorphism in this case) to flow the germ $\tilde{f}$ to any other germ $\tilde{g}$ with $j^k(\tilde{f}) = j^k(\tilde{g})$. Consider

$$H_t(x) = H(x,t) = (1 - t)\tilde{f}(x) + t\tilde{g}(x)\,,$$

so that $H(x,0) = \tilde{f}$ and $H(x,1) = \tilde{g}$. It turns out that it is equivalent to find a vector field $(\eta(x,t),1)$, where $\eta \in \varepsilon(n + 1,n)$, such that $(\eta,1)$ is tangent to the level surface H = constant. Let $\Phi_{(x_0,t_0)}(t)$ be the x-component of the trajectory of $(\eta,1)$, then $H(\Phi_{(x_0,t_0)},t)$ = constant for all t. Now differentiate this equation, we have

$$(\text{grad } H)\cdot\eta = 0\,. \tag{3.2}$$

With this analysis, we are required to find a criterion such that $(\text{grad } H) \cdot \eta = 0$. It turns out that this criterion is:

$$m(n)^{k+1} \subset m^2 \langle \frac{\partial \tilde{f}}{\partial x} \rangle . \tag{3.3}$$

In the process of finding this criterion we need the following important lemma which is due to Nakayama.

<u>Lemma 3.2</u>. (Nakayama) Let R be a commutative ring with identity. Let I be an ideal in R such that $1 + z$ is invertible for all $z \in I$. Let A, B be submodules of some R-module M and suppose A is finitely generated over R. Then $A \subseteq B + IA$ implies $A \subseteq B$.

(Note that the role of M is simply to ensure that $B + IA$ has meaning.)

<u>Proof of Lemma</u>: Let A be generated by $a_1, \dots, a_n$. Then

$$\begin{aligned} a_1 &= b_1 + \sum_{i=1}^{n} z_{1i} a_i \\ &- - - - - - - - - - - - \\ a_n &= b_n + \sum_{i=1}^{n} z_{ni} a_i \end{aligned} \tag{3.4}$$

for some $b_1, \dots, b_n \in B$ and $z_{ji} \in I$, $1 \leq j, i \leq n$. Then

$$\begin{aligned} b_1 &= (1 - z_{11})a_1 - z_{12}a_2 - \dots - z_{1n}a_n \\ &- \\ b_n &= -z_{n1}a_1 - \dots + (1 - z_{nn})a_n \end{aligned} \tag{3.5}$$

By Cramer's rule

$$a_j = \frac{\begin{vmatrix} (1 - z_{11}) \dots b_1 \dots - z_{1n} \\ - - - - - - - - - - - - - \\ -z_{n1} \dots b_n \dots (1 - z_{nn}) \end{vmatrix}}{\begin{vmatrix} 1 - z_{11} & \dots & - z_{1n} \\ & \ddots & \\ -z_{n1} & \dots & 1 - z_{nn} \end{vmatrix}} = \frac{\begin{vmatrix} (1 - z_{11}) \dots b_1 \dots - z_{1n} \\ - - - - - - - - - - - - - \\ -z_{n1} \dots b_n \dots (1 - z_{nn}) \end{vmatrix}}{1 + z} \tag{3.6}$$

for some $z \in I$ and $b_1, \ldots, b_n$ are occupying the jth column. Since $1 + z$ is invertible, we can write

$$a_j = \sum_{k=1}^{n} z_{jk} b_k \tag{3.7}$$

where $z_{jk} \in I$. Hence $A \subseteq B$.

Proof of Theorem 3.1: Let $\tilde{g} \in m(n)$ and suppose $j^k(\tilde{f}) = j^k(\tilde{g})$. Given $m(n)^{k+1} \subset m^2\langle\frac{\partial \tilde{f}}{\partial x}\rangle$, we wish to show that $\tilde{f} \sim \tilde{g}$. Construct the homotopy

$$H_t(x) = H(x,t) = (1 - t)f(x) + tg(x) , \tag{3.8}$$

for $0 \leq t \leq 1$. Then

$$\frac{\partial H}{\partial x_i}(x,t) - \frac{\partial f}{\partial x_i}(x) = t\left(\frac{\partial g}{\partial x_i}(x) - \frac{\partial f}{\partial x_i}(x)\right) . \tag{3.9}$$

Since $j^k(g) = j^k(f)$, $g - f \in m(n)^{k+1}$, and it follows that

$$\frac{\partial H}{\partial x_i} - \frac{\partial f}{\partial x_i} \in m(n)^k \varepsilon(n + 1) . \tag{3.10}$$

Now embed $\varepsilon(n)$, $m(n)$ in $\varepsilon(n + 1)$, $m(n + 1)$ in the obvious way. Then

$$m(n)^{k+1}\varepsilon(n + 1) \subseteq m(n)^2\langle\frac{\partial f}{\partial x}\rangle\varepsilon(n + 1) \tag{3.11}$$

$$\subseteq m(n)^2\langle\frac{\partial H}{\partial x}\rangle\varepsilon(n + 1) + m(n)^{k+2}\varepsilon(n + 1) \tag{3.12}$$

$$\subseteq m(n)^2\langle\frac{\partial H}{\partial x}\rangle\varepsilon(n + 1) + m(n + 1)m(n)^{k+1}\varepsilon(n + 1) \tag{3.13}$$

We apply the Nakayama Lemma with $R = \varepsilon(n + 1)$, $I = m(n + 1)$, $A = m(n)^{k+1}\varepsilon(n + 1)$, $B = m(n)^2\langle\frac{\partial H}{\partial x}\rangle\varepsilon(n + 1)$, to ensure that

$$m(n)^{k+1}\varepsilon(n + 1) \subseteq m(n)^2\langle\frac{\partial H}{\partial x}\rangle\varepsilon(n + 1) . \tag{3.14}$$

Now $\frac{\partial H}{\partial t} = g - f \in m(n)^{k+1} \subseteq m(n)^{k+1}\varepsilon(n + 1)$, hence $\frac{\partial H}{\partial t} \in m(n)^2\langle\frac{\partial H}{\partial x}\rangle\varepsilon(n + 1)$. Thus we can find $\mu_1, \ldots, \mu_n \in m(n)^2\varepsilon(n + 1)$ such that

$$\sum_{i=1}^{n} \frac{\partial H}{\partial x_i}(x,t)\mu_i(x,t) + \frac{\partial H}{\partial t}(x,t) = 0 \tag{3.15}$$

(This is how we find the vector field we mentioned before.) The differential equation

$$\frac{\partial \Phi}{\partial t}(x,t) = \mu(\Phi(x,t),t) , \tag{3.16}$$

where $\Phi \in \varepsilon(n+1,n)$, $\mu \in \varepsilon(n+1)$, has a unique solution near $t = t_0$ with the initial condition $\Phi_{t_0} = \mathrm{Id}$. Note, moreover, that $\Phi(\bar{0},t) = \bar{0}$ since $\mu(\bar{0},t) = 0$. From (3.15) and (3.16) it follows that

$$\frac{d}{dt}H(\Phi(x,t),t) = 0 . \tag{3.17}$$

Therefore the vector field we mentioned before is indeed tangent to the level surface H = constant. Then $H(\Phi(x_0,t_0),t)$ = constant for all t along the x-component of the trajectory of the vector field so that $H_t\Phi_t$ is independent of t near t_0 where $\Phi_t \in \varepsilon(n,n)$ is defined $\Phi_t(x) = \Phi(x,t)$. Thus

$$H_t\Phi_t = H_{t_0} ; \tag{3.18}$$

moreover Φ_t is a diffeomorphism since $\Phi_{t_0} = \mathrm{Id}$. Thus $H_t \sim H_{t_0}$ for t sufficiently close to t_0. Now using the compactness and connectedness of the closed interval $[0,1]$, we may cover $[0,1]$ be a finite number of neighborhoods $U_1,\ldots,U_\lambda$ such that, for any t_i, t_j in any one of these neighborhoods, then

$$H_{t_i} \sim H_{t_j} . \tag{3.19}$$

Hence

$$\tilde{f} = H_0 \sim H_{t_1} \sim \ldots \sim H_1 = \tilde{g} . \tag{3.20}$$

<u>Remark</u>. The converse of this theorem is not true. Let us look at the following counterexample [64] (due to D. Siersma): consider

$$f(x_1,x_2) = \frac{1}{3}x_1^3 + x_1x_2^3 = x_1(\frac{1}{3}x_1^2 + x_2^3) . \qquad (3.21)$$

By computation it is not hard to find that $\tilde{f}$ is 4-determined. However,

$$\langle\frac{\partial f}{\partial x}\rangle = \langle x_1^2 + x_2^3 , x_1x_2^2\rangle$$

and

$$m(2)^2 = \langle x_1^2, x_1x_2, x_2^2\rangle .$$

Therefore

$$m(2)^2\langle\frac{\partial f}{\partial x}\rangle = \langle x_1^4 + x_1^2x_2^3 , x_1^3x_2 + x_1x_2^4 , x_1^2x_2^2 + x_2^5 , x_1^3x_2^2 , x_1^2x_2^3 , x_1x_2^4\rangle .$$

$x_2^5 \notin m(2)^2\langle\frac{\partial f}{\partial x}\rangle$ implies that $m(2)^5 \not\subset m(2)^2\langle\frac{\partial f}{\partial x}\rangle$ although $\tilde{f}$ is not 4-determined. Since $x_1^6 = x_1^2(x_1^4 + x_1^2x_2^3) - x_1^2(x_1^2x_2^3)$, $x_1^5x_2 = x_1x_2(x_1^4 + x_1^2x_2^3) - x_1x_2(x_1^2x_2^3)$, $x_1^4x_2^2 = x_1(x_1^3x_2^2)$, etc., it is easy to see that $m(2)^6 \subset m(2)^2\langle\frac{\partial f}{\partial x}\rangle$. Thus, by Theorem 3.1, $\tilde{f}$ is 5-determined.

The necessary condition for a ferm to be k-determined will be given in the next theorem which will not be proved here since the references we provide give quite clear and straight forward proofs.

<u>Theorem 3.3</u>. If $\tilde{f} \in m(n)$ and $\tilde{f}$ is k-determined then $m(n)^{k+1} \subset m(n)\langle\frac{\partial f}{\partial x}\rangle$.

For a proof of the theorem, see [43,82]. The condition of this theorem is not a sufficient condition for k-determinancy. The following simple example shows that Theorem 3.3 is not reversible. Let $f(x) = x^2$. It is clear that $m(1)^2 = m(1)\langle\frac{df}{dx}\rangle$ and it is equally clear that $j^1(f)$ is not sufficient, so that $m(1)^2 \subset m(1)\langle\frac{df}{dx}\rangle$ although $\tilde{f}$ is not 1-determined.

Now, let us put Theorem 3.1 and Theorem 3.3 together and generate the following interesting corollary.

<u>Corollary 3.4</u>. $\tilde{f}$ is finitely determined if and only if $m(n)^k \subset \langle\frac{\partial f}{\partial x}\rangle$ for some k.

Proof: If $\tilde{f}$ is finitely determined, say k-determined then

$$m(n)^{k+1} \subset m(n)\langle\frac{\partial f}{\partial x}\rangle \subset \langle\frac{\partial f}{\partial x}\rangle . \qquad (3.22)$$

If there is k such that

$$m(n)^{k} \subset \langle\frac{\partial f}{\partial x}\rangle , \qquad (3.23)$$

and

$$m(n)^{k+2} \subset m(n)^2\langle\frac{\partial f}{\partial x}\rangle . \qquad (3.24)$$

By Theorem 3.1, $\tilde{f}$ is $(k+1)$-determined.

Remark. Suppose $\tilde{f} \in m(n)$ and $\tilde{f}$ is $r\ell$ k-determined (k being finite), then $\tilde{f}$ is $(k+2)$-determined. (For a proof, see [82, p. 44]). Therefore $\tilde{f}$ is right finitely determined if and only if it is right-left finite-determined. We may omit the modifier right and right-left for finite determinancy (but not for k-determinancy).

Definition 3.1. If $\tilde{f} \in m(n)^2$, the codimension of $\tilde{f}$, denoted by $\text{cod}\,\tilde{f}$, is defined to be the integer $\dim_{\mathbb{R}} m(n)/\langle\frac{\partial f}{\partial x}\rangle$.

This definition makes sense because if $\tilde{f} \in m(n)^2$, each $\frac{\partial f}{\partial x_i} \in m(n)$ and so $\langle\frac{\partial f}{\partial x}\rangle \in m(n)$. In fact, this definition can also be justified from the geometrical point of view; namely $\text{cod}\,\tilde{f}$ is indeed the codimension of the orbit of $\tilde{f}$ under the group of diffeomorphisms $L(n)$ in the space $m(n)^2/m(n)^{k+1}$ in case $0 \leq \text{cod}\,\tilde{f} \leq k-2$. (For a proof see either [43] or [82].)

Moreover, $\text{cod}\,\tilde{f}$ should really be denoted as right-codimension of $\tilde{f}$ or $r\text{-cod}\,\tilde{f}$. In order to define right-left $\text{cod}\,\tilde{f}$ we introduce the canonical induced $\mathbb{R}$-algebra homomorphism $g^*: \varepsilon(p) \to \varepsilon(n)$, where $\tilde{g} \in \varepsilon(n,p)$ and $g^*(\tilde{f}) = f \circ g$ if $\tilde{f} \in \varepsilon(p)$.

Definition 3.2. Let $\tilde{f} \in m(n)^2$. The right-left codimension of $\tilde{f}$, $r\ell\text{-cod}\,\tilde{f}$, is $\dim_{\mathbb{R}} m(n)/(\langle\frac{\partial f}{\partial x}\rangle + f^* m(1))$.

Remark. For the purposes of this book we are mostly interested in the case when the codimension of a germ is less than or equal to 4. With this restriction, we have that $\text{cod}\,\tilde{f}$ is the same as the right-left $\text{cod}\,\tilde{f}$. As a matter of fact most of our definitions and theorems will be stated in the right case, therefore we will omit the modifier "right" for terms we will use unless we wish to indicate specifically. We choose to do so because we feel that it will then be easier for the reader to follow the main stream of the argument (avoiding some technical details), although we will, in principle, thereby lose a class of germs in the discussion. For the details of right-left case, we recommend the reader to Wasserman's "Stability of Unfoldings" [82].

By abuse of language we will call a set of germs in $m(n)$ a <u>basis</u> for $m(n)/\langle\frac{\partial f}{\partial x}\rangle$ if the projections of the elements of the set onto $m(n)/\langle\frac{\partial f}{\partial x}\rangle$ constitute an $\mathbb{R}$ basis.

<u>Example 3.2</u>. $\tilde{f} = x^k \in m(1)$, $\operatorname{cod} \tilde{f} = \dim_{\mathbb{R}} \langle x\rangle/\langle x^{k-1}\rangle = k - 2$ with basis $\{x, x^2, \ldots, x^{k-2}\}$.

<u>Example 3.3</u>. $\tilde{f} = x^2y \in m(2)$, $\langle\frac{\partial f}{\partial x}\rangle = \langle 2xy, x^2\rangle$. In this case, the cosets of $y, y^2, y^3, \ldots$ will never be zero. Thus $\operatorname{cod} \tilde{f} = \infty = \dim_{\mathbb{R}} \langle x,y\rangle/\langle 2xy, x^2\rangle$.

<u>Example 3.4</u>. $\tilde{f} = x^2y \pm y^k \in m(2)$ is k-determined and $\operatorname{cod} \tilde{f} = k$. A suitable basis for $m(2)/\langle xy, x^2 \pm y^{k-1}\rangle$ is provided by $\{x, y, y^2, \ldots, y^{k-1}\}$.

<u>Example 3.5</u>. $\tilde{f} = x^3 + y^3 \in m(2)$, $\operatorname{cod} \tilde{f} = \dim_{\mathbb{R}} \langle x,y\rangle/\langle x^2, y^2\rangle = 3$ with basis $\{x, y, xy\}$.

<u>Example 3.6</u>. $\tilde{f} = x^4 + y^4 \in m(2)$, $\operatorname{cod} \tilde{f} = \dim_{\mathbb{R}} \langle x,y\rangle/\langle x^3, y^3\rangle = 8$ with basis $\{x, y, x^2, xy, y^2, x^2y, xy^2, x^2y^2\}$.

<u>Example 3.7</u>. $\tilde{f} = x^3 + y^3 + z^3 \in m(3)$, $\operatorname{cod} \tilde{f} = \dim_{\mathbb{R}} \langle x,y,z\rangle/\langle x^2, y^2, z^2\rangle = 7$ with basis $\{x, y, z, xy, xz, yz, xyz\}$.

4. <u>Universal Unfoldings</u>

In this section we shall not distinguish a germ and its representative unless we encounter possible confusion.

<u>Definition 4.1</u>. Let $f \in m(n)$. A germ of $F \in m(n + r)$ is called an

<u>unfolding</u> of f if $F|_{\mathbb{R}^n \times \{0\}} = f$, i.e. $F(x,0) = f(x)$, $x \in \mathbb{R}^n$.

We may write $F_u(x) = F(x,u)$, where $x \in \mathbb{R}^n$, $u \in \mathbb{R}^r$, and think of the unfolding F as an r-parameter family of germs. We call r the <u>codimension</u> of the unfolding F; we may write (F,r) for F to emphasize the codimension.

<u>Example 4.1</u>. For any $f \in m(n)$, the <u>constant unfolding</u> F of codimension r is defined by

$$F(x,u) = f(x), \qquad x \in \mathbb{R}^n, \qquad u \in \mathbb{R}^r.$$

<u>Example 4.2</u>. Let $b \in m(n)$, $u = (u_1,\ldots,u_r) \in \mathbb{R}^r$. Then

$$F(x,u_1) = f(x) + b(x)u_1$$

is an unfolding of $f \in m(n)$ of codimension 1. More generally, let $b_1,\ldots,b_r \in m(n)$. Then

$$G(x,u) = f(x) + \sum_{i=1}^{r} b_i(x)u_i$$

is an unfolding of $f \in m(n)$ of codimension r.

This example suggests implicitly the notion of the <u>sum</u> of two unfoldings of $f(x)$.

<u>Definition 4.2</u>. Let (F,r) and (G,s) be two unfoldings of $f \in m(n)$. We define the <u>sum</u> of (F,r) and (G,s) to be

$$(F,r) + (G,s) = (H,r+s)$$

where

$$H(x,u,v) = F(x,u) + G(x,v) - f(x) \tag{4.1}$$

where $x \in \mathbb{R}^n$, $u \in \mathbb{R}^r$, $v \in \mathbb{R}^s$.

We then see (Example 4.2) that $F(\overline{x},\overline{u}) = f(\overline{x}) + \Sigma\, b_i(\overline{x})u_i$ is the sum of the codimension 1 unfoldings $f(\overline{x}) + b_i(\overline{x})u_i$. In general the codimension is additive with respect to the sum of unfoldings.

Example 4.3. Let $f \in m(1)$ be defined by $f(x) = x^3$. $F(x,u) = x^3 + u^2$ is an unfolding of f. $G(x,u) = x^3 + ux$ is also an unfolding of f. Intuitively G will give us all the deformations of f as we vary the parameter u. This is the naive view point of versal unfolding, which will be defined later. Similarly $H(x,w,v) = x^3 + wx + vx$ and $K(x,v) = x^3 + 3vx^2$ are also unfoldings of f.

A natural question has been raised by our second example: How do we distinguish (or associate) two unfoldings? We begin with an informal discussion. Let $f \in m(n)$. Let (F,r) and (G,s) be two unfoldings of f. Then F and G are said to be (right) associated if there exists a C^∞ map $\Phi: (\mathbb{R}^n \times \mathbb{R}^s, \overline{0}) \to (\mathbb{R}^n \times \mathbb{R}^r, \overline{0})$, where $\overline{0} \in \mathbb{R}^n$, having the following properties:

(1) $\Phi|_{\mathbb{R}^n \times \{\overline{0}\}} = \mathrm{id}|_{\mathbb{R}^n}$ i.e. $\Phi(x,\overline{0}) = (x,\overline{0})$

(2) Φ is fibrewise in the sense that if we keep $\overline{v} \in \mathbb{R}^s$ constant, then Φ will map $\mathbb{R}^n \times \{\overline{v}\}$ into some $\mathbb{R}^n \times \{u\}$, where $\overline{u} \in \mathbb{R}^r$ depends only on $\overline{v}$. We require this map to be smooth.

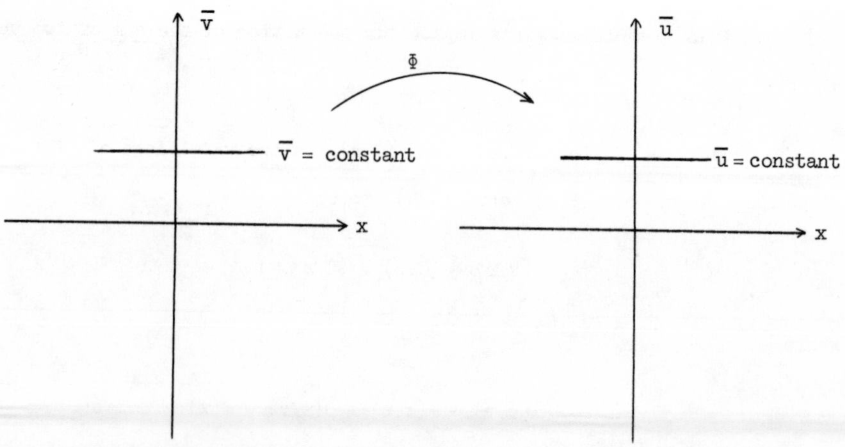

We write $\overline{u} = \psi(\overline{v})$.

(3) The graph of G on $\mathbb{R}^n \times \{\overline{v}\}$ when $\overline{v}$ = constant and the graph of F

on $\mathbb{R}^n \times \{\bar{u}\}$ when $\bar{u}$ = constant, where $\bar{u} = \psi(\bar{v})$ differ only by a translation. These translations may depend on $\bar{v}$, i.e. we could have different translations for different $\bar{v}$'s, this mapping on $\mathbb{R}^s$ is also required to be C^∞.

Example 4.4. Referring back to Example 4.3,

(i) $G(x,u) = x^3 + ux$ is associated to $H(x,w,v) = x^3 + wx + vx$. Define $\Phi: \mathbb{R}^1 \times \mathbb{R}^2 \to \mathbb{R}^1 \times \mathbb{R}^1$ by $\Phi(x,w,v) = (x,u)$, where $u = w + v$. In this case, if we keep w, v constant, we get a line parallel to the x-axis in (x,w,v)-space, which will be mapped into a line in (x,u) space parallel to the x-axis. Then the graph of H for w, v being kept constant is the same as the graph of G where $u = w + v$. Notice that we do not need any translation in this case.

(ii) $G(x,u)$ is associated to $K(x,v) = x^3 + 3vx^2$. Define $\Phi: \mathbb{R}^1 \times \mathbb{R}^1 \to \mathbb{R}^1 \times \mathbb{R}^1$ by $\Phi(x,v) = (x + v,u)$. Since $K(x,v) = x^3 + 3vx^2 = (x + v)^3 - 3v^2(x + v) + 2v^3$, it is clear that u should be defined as $-3v^2$ so that $K(x,v) = G\Phi(x,v) + 2v^3$ and so, for each v, we need a translation

$$\lambda: \mathbb{R}^1 \times \mathbb{R}^1 \to \mathbb{R}$$

given by

$$\lambda(t,v) = t - 2v^3 .$$

Otherwise written,

$$\lambda(t,v) = t + \alpha(v)$$

where $\alpha(v) = -2v^3$.

Now the formal definition of association of two unfoldings (F,r) and (G,s) of a germ $f \in m(n)$ may be given as follows.

Definition 4.3. (F,r) and (G,s) are associated if there exists a morphism $(\Phi,\psi,\lambda)\colon (G,s) \to (F,r)$, where Φ is a germ: $(\mathbb{R}^{n+s},\overline{0}) \to (\mathbb{R}^{n+r},\overline{0})$, ψ is a germ $(\mathbb{R}^{s},\overline{0}) \to (\mathbb{R}^{r},\overline{0})$ and a translation $\lambda\colon (\mathbb{R}^{1+s},\overline{0}) \to (\mathbb{R},0)$ with $\lambda(t,v) = t + \alpha(v)$ where $\alpha\colon (\mathbb{R}^{s},\overline{0}) \to (\mathbb{R},0)$ such that

1. $\Phi|_{\mathbb{R}^n \times \{\overline{0}\}} = \mathrm{id}|_{\mathbb{R}^n}$ i.e. $\Phi(x,\overline{0}) = (x,\overline{0})\ \forall\ x \in \mathbb{R}^n$.
2. $\pi_r\Phi = \psi\pi_s$ where π_r, π_s are natural projections.
3. $G = \lambda(F\Phi,\pi_s) = F\Phi + \alpha\pi_s$.

Diagramatically, we have the following:

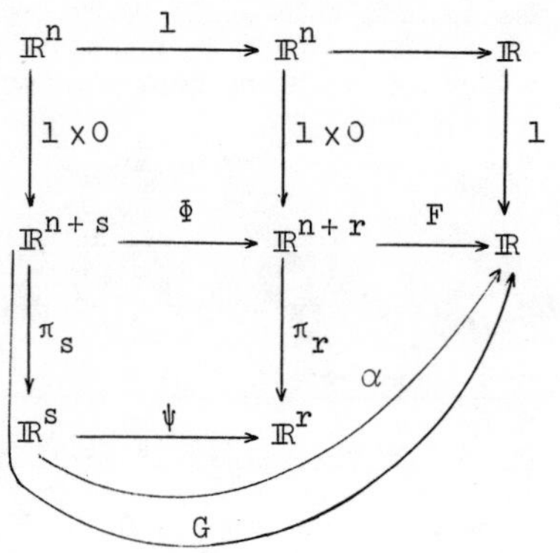

Sometimes we will denote a morphism (Φ,ψ,λ) simply by (Φ,ψ,α).

In order to be consistent with the notion of the (right) equivalence of germs, we should require $\lambda_v \in \epsilon(1)$, $\lambda_v(t) = \lambda(t,v)$, to be the identity mapping for each v instead of allowing λ to be a translation. However there is no essential difference. Just as we did in the case of jet extensions, we are in effect ignoring the constant term. More precisely we are here ignoring the constant term of the restrictions of G to the fibres. This is reasonable since we are interested in unfoldings as families of germs in $m(n)$ and the constant term plays no role in this aspect.

It would be possible to replace these translations by more general transformations:

Definition 4.4. (F,r) and (G,s) are right-left associated if there exists a right-left morphism (Φ,ψ,λ) where Φ, ψ are the same as before, but

$\lambda \in \epsilon(1+s)$, $\lambda|_{\mathbb{R}} = \mathrm{id}|_{\mathbb{R}}$, and $G = \lambda(F\Phi, \pi_s)$.

Note that the map λ mentioned in the definition is not required to be a translation in contrast to the roll of λ in Definition 4.3 .

Now given a morphism $(\Phi_1, \psi_1, \lambda_1): (F_1, r_1) \to (F, r)$ and a morphism $(\Phi_2, \psi_2, \lambda_2): (F_2, r_2) \to (F_1, r_1)$, we may compose these morphisms as follows.

$$(\Phi_2, \psi_2, \lambda_2) \circ (\Phi_1, \psi_1, \lambda_1) = (\Phi, \psi, \lambda) , \tag{4.2}$$

where $\Phi = \Phi_2 \circ \Phi_1$, $\psi = \psi_2 \circ \psi_1$ and $\alpha = \alpha_1 \psi_2 + \alpha_2$ with $\lambda_1(t,v) = t + \alpha_1(v)$, $\lambda_2(t,u) = t + \alpha_2(u)$ and $\lambda(t,w) = t + \alpha(w)$. Clearly (Φ, ψ, λ) is a morphism $(F_2, r_2) \to (F, r)$. In this way we obtain a <u>category</u> of unfoldings of f . For the <u>right-left category</u>, the only difference is that we must describe λ by the more general rule

$$\lambda(t,w) = \lambda_2(\lambda_1(t, \psi_2(w)), w) \tag{4.3}$$

for $w \in \mathbb{R}^{r_2}$.

When should two unfoldings be regarded as the same? We answer this question in the following definition.

<u>Definition 4.5</u>. The two unfoldings (F,r) and (G,s) of $f \in m(n)$ are <u>equivalent</u> (or <u>isomorphic</u>) if they are equivalent objects in the category of unfoldings of f .

<u>Remarks</u>. (1) Geometrically, (F,r) is equivalent to (G,s) if

(i) $r = s$;

(ii) three is a morphism (Φ, ψ, α) from (G,s) to (F,r) , where ψ is a reparametrization of $\mathbb{R}^r$ (or $\psi \in L(r)$) and $\Phi_u: \mathbb{R}^n \to \mathbb{R}^n$ is a reparametrization of $\mathbb{R}^n$. Thus (Φ, ψ, α) has an inverse $(\Phi^{-1}, \psi^{-1}, -\alpha\, \Phi^{-1})$.

(2) The <u>$r\ell$-equivalence</u> of two unfoldings can be similarly defined as in Definition 4.5 .

Definition 4.6. An unfolding (F,r) of $f \in m(n)$ is versal (or stable) if for any unfolding (G,s) of f, there is a morphism (Φ,ψ,α) from (G,s) to (F,r).

Definition 4.7. An unfolding (F,r) is universal if it is stable and r is the minimal codimension of F.

From these definitions we see that a universal unfolding is a particular case of the more general notion of a stable unfolding. Now, in the theory of singularities or structurally stable mappings, as soon as one formulates a notion of stability, the question is raised of discovering criteria for a map to be stable. Thus, Mather's notion of infinitesmal stable maps arose as an attempt to algebrize the analytical definition of stability. The role that transversality plays is as a bridge between stability and infinitesmal stability enabling one to work purely in a finite codimensional setting. As a result, it is a very important concept to which we now direct our attention. Since it is by no means easy to understand, we again use an example to provide an intuitive idea behind the definition of transversality.

Consider $J^3(1,1)$. Decompose $J^3(1,1) \cong \mathbb{R}^3 = M_0 \cup M_1 \cup M_2 \cup M_3$, where $M_0 = \{(0,0,0)\}$, $M_1 = \{\text{c-axis} - (0,0,0)\} = M_1^+ \cup M_1^-$ where M_1^+ is the positive c-axis and M_1^- the negative c-axis, $M_2 = \{\text{plane - c-axis}\}$ and $M_3 = \mathbb{R}^3 - \{(a,c) \text{ plane}\}$ as in Figure 4.1.

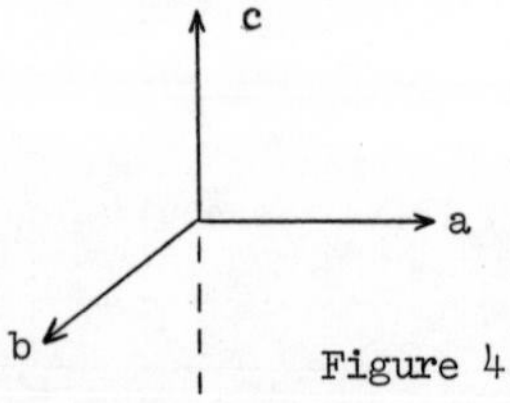

Figure 4.1.

Let $f(x) = x^3$ be a germ in $m(1)$ and let $F(x,u) = x^3 + ux$ be an unfolding of f. We are about to see how to assign a meaning to the statement that F is the universal unfolding of f. As we explained before J^3f maps (or throws) $\mathbb{R}^1$ into $J^3(1,1)$ as a parabola with vertex at $(0,0,1)$ since

$$j^3(f)(t) = x^3 + 3tx^2 + 3t^2x \,. \tag{4.4}$$

Thus $a = 3t^2$, $b = 3t$ and $c = 1$, equivalently we have the parabola $b^2 = 3a$ at $c = 1$ as Figure 4.2 .

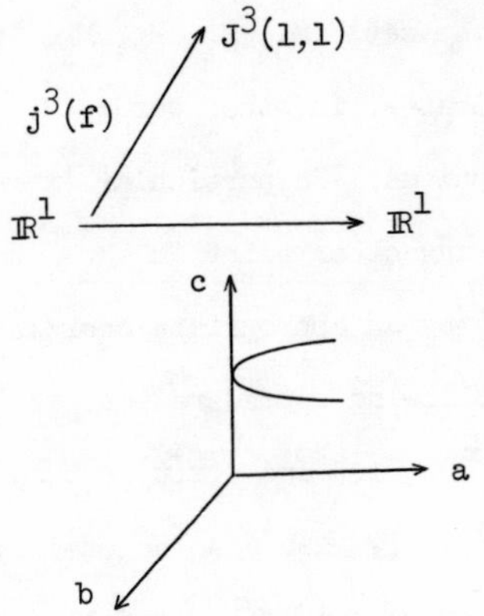

Figure 4.2

It is quite clear that this parabola is not transversal to the c-axis since if we perturb the map a little, to $f + \delta$, then $J^3(f + \delta)(\mathbb{R}^1)$ will most likely miss the c-axis as Figure 4.3 .

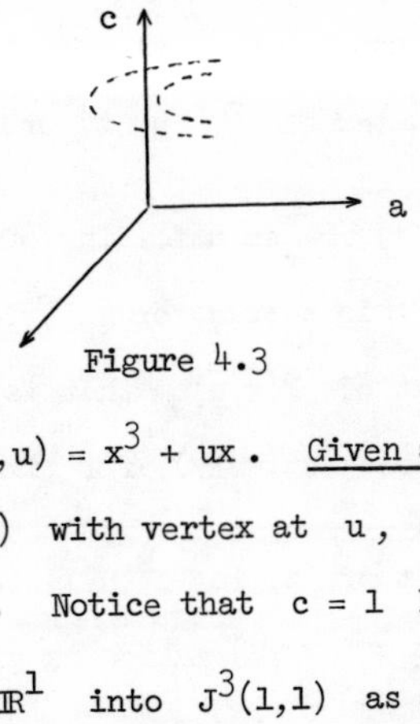

Figure 4.3

Now, let us consider $F(x,u) = x^3 + ux$. <u>Given a fixed u</u> , $J^3(F)(\mathbb{R})$ gives another parabola $b^2 = 3(a - u)$ with vertex at u , since $F_u'(x) = 3x^2 + u$, $\frac{1}{2!} F_u''(x) = 3x$, $\frac{1}{3!} F_u'''(x) = 1$. Notice that $c = 1$ has not been changed, which tells us that $J^3(F)$ throws $\mathbb{R}^1$ into $J^3(1,1)$ as a parabola in the plane $c = 1$ with vertex at u . Let us perform this process for all u , then the (x,u) plane (or $\mathbb{R}^1 \times \mathbb{R}^1$) will be thrown onto the $\{(a,b,c) | c = 1\}$ plane in $J^3(1,1)$. Intuitively, what we are doing is to take the union of all parabolas which have been thrown into $J^3(1,1)$ by $J^3(F)$ for some fixed u . Observe that the resulting plane is transversal to M_1 . In other words, the map F , which maps the (x,u) space into the (a,b,c) space is transversal to the c-axis (the only point of intersection is the point $(0,0,1)$). This is indeed the concept behind the statement that F universally unfolds f.

Notice that in each of the sets M_0, M_1, M_2, M_3, jets can be transformed into each other by changing coordinates, in other words M_i is the orbit of a jet under the group of diffeomorphisms. In particular x^3 in $J^3(1,1)$ represents a point $(0,0,1)$ in M_1, and any other point in M_1, say $(0,0,c)$ can certainly be transformed into $(0,0,1)$ by an appropriate coordinate change.

Formally, we have the following consideration.

Let $F\colon U(\text{open in } \mathbb{R}^{n+r}) \to \mathbb{R}$ and $(x,u) \in U$. Define $F_{(x,u)}$ to be the germ in $m(n)$ of the representative map also denoted by $F_{(x,u)}$ which is defined on a small open neighborhood of x in $\mathbb{R}^n$ as follows:

$$F_{(x,u)}(y) = F(x+y,u) - F(x,u).$$

Then, let $\hat{F} = J^kF\colon U \to J^k(n,1)$ be the jet extension of F defined by $\hat{F}(x,u) = J^kF(x,u) = \pi_k(F_{(x,u)})$. Here $\pi_k\colon \varepsilon(n,1) \to J^k(n,1) = \varepsilon(n,1)/m(n)^{k+1}$ is the natural projection map and a canonical element of the coset of $F_{(x,u)}$ is its Taylor expansion, evaluated at $\overline{0}$, up to order k.

Definition 4.8. Let (F,r) be an unfolding of $f \in m(n)$. Let $z = \pi_k(f) \in J^k(n,1)$. Then F is k-transversal (right-left k-transversal, or $r\ell$ k-transversal) if the k-jet of $F(\overline{x},\overline{u})$, that is, $\hat{F}(\overline{x},\overline{u})$ is transversal at $\overline{0}$ to $zL^k(n)$ $(L^k(1)zL^k(n)$ respectively), for fixed $\overline{u}$.

Here $zL^k(n)$ is the orbit of z in $J^k(n,1)$ under the group of k-jets of diffeomorphisms which fix $\overline{0} \in \mathbb{R}^n$. The action is given by $z \cdot \mu = \pi_k(f \circ g)$ where $g\colon (\mathbb{R}^n, \overline{0}) \to (\mathbb{R}^n, \overline{0})$ is a local diffeomorphism of $\mathbb{R}^n$ with $j^k(g) = \mu$.

Example 4.5. Let $f(x) = x^4$ be in $m(1)$. In particular $f \in J^4(1,1) \cong \mathbb{R}^4$. Since

$$f(x) = x^4 = (x-t)^4 + 4(x-t)^3t + 6(x-t)^2t^2 + 4(x-t)t^3 + t^4, \qquad (4.5)$$

the jet-extension $J^4f(t)$, of f, is $x^4 + 4tx^3 + 6t^2x^2 + 4t^3x$. In $J^4(1,1)$,

$J^4 f(t)$ is the point $(4t^3, 6t^2, 4t, 1)$. Equivalently it represents a point (a,b,c,d) in $\mathbb{R}^4$ where $a = 4t^3$, $b = 6t^2$, $c = 4t$ and $d = 1$. This is a twisted cubic which is not transversal to the d-axis at $d = 1$. However if we consider

$$F(x,u,v) = x^4 + ux^2 + vx , \tag{4.6}$$

by adding the two parameters u and v, then fixing each u, v, we get a similar curve. Taking the union of all the resulting curves as u and v vary, the resulting surface is a three-dimensional surface which is transversal to the d-axis at $d = 1$. Thus we say that F is 4-transversal.

Let F unfold f, $F \in \varepsilon(n + r)$. Let $\alpha_i(F) = \frac{\partial F}{\partial u_i}\Big|_{(\mathbb{R}^n, \bar{0})} \in \varepsilon(n)$, $i = 1,\ldots,r$, where $\{u_1,\ldots,u_r\}$ is the coordinate system of $\mathbb{R}^r$. Let $V_F = \{\sum_{i=1}^{r} a_i\alpha_i(F) + a_0 1 : a_0, a_1,\ldots,a_r \in \mathbb{R}\}$ and $W_F = \{\sum_{i=1}^{r} a_i\alpha_i(F) : a_1,\ldots,a_r \in \mathbb{R}\}$.

Theorem 4.1.

(i) An unfolding (F,r) of f is k-transversal if and only if

$$\varepsilon(n) = \langle \frac{\partial f}{\partial x_1},\ldots,\frac{\partial f}{\partial x_n}\rangle_{\varepsilon(n)} + V_F + m(n)^{k+1} .$$

(ii) F is $r\ell$ k-transversal if and only if

$$\varepsilon(n) = \langle \frac{\partial f}{\partial x_1},\ldots,\frac{\partial f}{\partial x_n}\rangle_{\varepsilon(n)} + f^*\varepsilon(1) + W_F + m(n)^{k+1} .$$

(In practice to find such a transversal F, we reduce modolo $m(n)^{k+1}$ to convert to a finite dimensional problem.)

To prove the theorem, we need the following lemma which is essentially a reinterpretation of the tangent space at z of $zL^k(n)$ or $L^k(1)zL^k(n)$ in $J^k(n,1)$. We refer the reader to [82] for the proof of this lemma.

Lemma 4.2. Let $f \in m(n)$, let k be a non-negative integer, and let $z = \pi_k(f) \in J^k(n,1)$. Then

(a) $\pi_k^{-1}[T_z(zL^k(n))] = m(n)\langle \frac{\partial f}{\partial x_1},\ldots,\frac{\partial f}{\partial x_n}\rangle + m(n)^{k+1}$

and

$$\text{(b)}\quad \pi_k^{-1}[T_z(L^k(1)zL^k(n))] = m(n)\langle\frac{\partial f}{\partial x_1},\ldots,\frac{\partial f}{\partial x_n}\rangle + f^*m(1) + m(n)^{k+1} .$$

Now to prove Theorem 4.1(i), we need to show that $d\hat{F}_{(\overline{0},\overline{0})}(T_{(\overline{0},\overline{0})}\mathbb{R}^{n+r}) + T_z(z \cdot L^k(n)) = J^k(n,1)$. Thus we need to find the differential of $\hat{F}$ at $\overline{0}$:

$$d\hat{F}_{(\overline{0},\overline{0})} = \left(\frac{\partial \hat{F}_i}{\partial x_j}\right)(\overline{0},\overline{0})$$

where $1 \leq i, j \leq n$. Now

$$\frac{\partial \hat{F}}{\partial x_j}(\overline{0},\overline{0}) = \frac{\partial}{\partial x_j}\pi_k(F_{(\overline{x},\overline{u})}(y))\Big|_{(\overline{x},\overline{u}) = (\overline{0},\overline{0})} = \pi_k[\frac{\partial}{\partial x_j}(F_{(\overline{x},\overline{u})}(y))\Big|_{(\overline{x},\overline{u}) = (\overline{0},\overline{0})}] .$$

Since

$$\frac{\partial}{\partial x_j} F_{(\overline{x},\overline{u})}(\overline{y})\Big|_{(\overline{x},\overline{u}) = (\overline{0},\overline{0})} = \frac{\partial}{\partial x_j}(F(\overline{y},\overline{0}) - F(\overline{0},\overline{0})) =$$

$$= \frac{\partial F}{\partial x_j}(\overline{y},\overline{0}) - \frac{\partial F}{\partial x_j}(\overline{0},\overline{0}) = \frac{\partial f}{\partial x_j}(\overline{y}) - \frac{\partial f}{\partial x_j}(\overline{0}) ,$$

$$\pi_k[\frac{\partial}{\partial x_j}(F_{(\overline{x},\overline{u})}(y))\Big|_{(\overline{x},\overline{u}) = (\overline{0},\overline{0})}] = \pi_k(\frac{\partial f}{\partial x_j}(\overline{y}) - \frac{\partial f}{\partial x_j}(\overline{0})) = \pi_k(\frac{\partial f}{\partial x_j}(\overline{y})) .$$

Similarly

$$\pi_k[\frac{\partial}{\partial u_\alpha}(F_{(\overline{x},\overline{u})}(\overline{y}))\Big|_{(\overline{x},\overline{u}) = (\overline{0},\overline{0})}] = \pi_k[\frac{\partial}{\partial u_\alpha}(F(\overline{0} + \overline{y},\overline{0}) - F(\overline{0},\overline{0}))]$$

$$= \pi_k[\frac{\partial F}{\partial u_\alpha}(\overline{y},\overline{0}) - \frac{\partial F}{\partial u_\alpha}(\overline{0},\overline{0})] = \pi_k(\frac{\partial F}{\partial u_\alpha}(\overline{y},\overline{0})) .$$

So, $d\hat{F}(\overline{0},\overline{0})(T_{(\overline{0},\overline{0})}\mathbb{R}^{n+r})$ is spanned by

$$\{\pi_k \frac{\partial f}{\partial x_1}(\overline{y}),\ \ldots,\ \pi_k \frac{\partial f}{\partial x_n}(\overline{y}),\ \pi_k \frac{\partial F}{\partial u_1}(\overline{y},\overline{0}),\ \ldots,\ \pi_k \frac{\partial F}{\partial u_r}(\overline{y},\overline{0})\}$$

over $\mathbb{R}$. Hence $\hat{F}$ is transversal to $zL^k(n)$ at $\overline{0}$ if and only if

$$\pi_k(m(n)\langle\frac{\partial f}{\partial x_1},\ldots,\frac{\partial f}{\partial x_n}\rangle) + \langle 1, \pi_k(\frac{\partial f}{\partial x_1}), \ldots, \pi_k(\frac{\partial f}{\partial x_n}), \pi_k(\frac{\partial F}{\partial u_1}), \ldots, \pi_k(\frac{\partial F}{\partial u_k})\rangle_{\mathbb{R}} = J^k(n,1)$$

or equivalently,

$$\varepsilon(n) = \langle\frac{\partial f}{\partial x_1}, \ldots, \frac{\partial f}{\partial x_n}\rangle_{m(n)} + \langle 1, \frac{\partial F}{\partial u_1}, \ldots, \frac{\partial F}{\partial x_n}\rangle_{\mathbb{R}} + m(n)^{k+1}$$

since

$$\langle\frac{\partial f}{\partial x_1}, \ldots, \frac{\partial f}{\partial x_n}\rangle_{\mathbb{R}} \subseteq \langle\frac{\partial f}{\partial x_1}, \ldots, \frac{\partial f}{\partial x_n}\rangle_{m(n)} .$$

A similar argument proves Theorem 4.1(ii).

In the light of this theorem, we have the following obvious corollaries.

<u>Corollary 4.3</u>. The unfolding (F,r) of f is k-transversal if $m(n) \subseteq \langle\frac{\partial f}{\partial x}\rangle + V_F + m(n)^{k+1}$, is right-left k-transversal if $m(n) \subseteq \langle\frac{\partial f}{\partial x}\rangle + V_F + m(n)^{k+1} + f^* m(1)$.

<u>Corollary 4.4</u>. Let $b_1, b_2, \ldots, b_r$ be elements of $m(n)$ which project to a basis for $m(n)/\langle\frac{\partial f}{\partial x}\rangle + m(n)^{k+1}$. Then the unfolding $F(x,u) = f(x) + \sum_{i=1}^{r} b_i(x)u_i$, where $u = (u_1, \ldots, u_r)$, is k-transversal.

<u>Corollary 4.5</u>. The sum of any unfolding and a k-transversal unfolding is k-transversal.

It is also as a consequence of Theorem 4.1 that we have the following two important results.

<u>Theorem 4.6</u>. Let $f \in m(n)$. Then for every non-negative integer k, there is an unfolding F of f which is k-transversal (and hence $r\ell$ k-transversal).

<u>Proof</u>: Let k be a non-negative integer. Since $\varepsilon(n)/(\langle\frac{\partial f}{\partial x}\rangle + m(n)^{k+1})$ is a finite dimensional vector space over $\mathbb{R}$, we may choose $b_1, \ldots, b_r \in \varepsilon(n)$ whose cosets modulo $\langle\frac{\partial f}{\partial x}\rangle + m(n)^{k+1}$ generate this vector space over $\mathbb{R}$. Define $F \in \varepsilon(n+r)$ by $F(x,u) = f(x) + \sum_{i=1}^{r} u_i b_i(x)$. Then $\frac{\partial F}{\partial u_i}\Big|_{\mathbb{R}^n \times \bar{0}} = b_i$ for

$1 \leq i \leq r$. Hence the b_i generate $W_F \subseteq V_F$. From Theorem 4.1 it is clear that F is k-transversal, furthermore it is also immediate from Theorem 4.1 that any k-transversal unfolding of f is also right-left k-transversal.

Theorem 4.7. For $f \in m(n)^2$, let (F,r) be a versal unfolding of f. Then F is k-transversal for any non-negative integer k.

Proof: From Theorem 4.6 there is certainly an unfolding (G,s) of f which is k-transversal. Since (F,r) is versal, there is a morphism $(\Phi,\psi,\lambda)\colon (G,s) \to (F,r)$ such that $G = F\Phi + \alpha\pi_s$ where $\lambda(t,\bar{v}) = t + \alpha(\bar{v})$. Hence $\frac{\partial\lambda}{\partial v_\ell}(t,\bar{0}) = \frac{\partial\alpha}{\partial v_\ell}(\bar{0})$ and is constant as a function of t. So $\frac{\partial\lambda}{\partial v_\ell}(t,0)$ is a constant germ, i.e. a real multiple of $1 \in \varepsilon(n)$, and hence $V_G = V_{F\Phi}$. We then claim that

$$V_{F\Phi} \subseteq \langle\frac{\partial f}{\partial x}\rangle + V_F . \tag{4.7}$$

If we can show (4.7), then, since $V_G = V_{F\Phi}$, we have

$$m(n) \subseteq \langle\frac{\partial f}{\partial x}\rangle + V_{F\Phi} + m(n)^{k+1} \subseteq \langle\frac{\partial f}{\partial x}\rangle + V_F + m(n)^{k+1} ,$$

and so (F,r) is k-transversal. Thus it remains to prove (4.7). Suppose $\Phi\colon (\bar{x},\bar{v}) \to (\bar{y},\bar{u})$ where $\bar{x},\bar{y} \in \mathbb{R}^n$, $\bar{u} \in \mathbb{R}^r$, $\bar{v} \in \mathbb{R}^s$. Then, since $\Phi(\bar{x},\bar{0}) = (\bar{x},\bar{0})$,

$$\frac{\partial F\Phi}{\partial v_\ell}(\bar{x},\bar{0}) = \sum_{i=1}^{n} \frac{\partial y_i}{\partial x_i}(\bar{x},\bar{0})\,\frac{\partial y_i}{\partial v_\ell}(\bar{x},\bar{0}) + \sum_{k=1}^{r} \frac{\partial F}{\partial u_k}(\bar{x},\bar{0})\,\frac{\partial u_k}{\partial v_\ell}(\bar{x},\bar{0}) . \tag{4.8}$$

We further notice that u_k does not depend on $\bar{x}$, that $\frac{\partial F}{\partial x_i}(\bar{x},\bar{0}) = \frac{\partial f}{\partial x_i}(\bar{x})$ and that $f \in m(u)^2$. Thus we have

$$\begin{aligned} \frac{\partial F\Phi}{\partial v_\ell}\Big|_{\mathbb{R}^n\times\bar{0}} - \frac{\partial F\Phi}{\partial v_\ell}(\bar{0},\bar{0}) &= \sum_{i=1}^{n} \frac{\partial f}{\partial x_i}\,\frac{\partial y_i}{\partial v_\ell}(\bar{x},\bar{0}) + \\ &+ \sum_{k=1}^{r} \Big(\frac{\partial F}{\partial u_k}\Big|_{\mathbb{R}^n\times\bar{0}} - \frac{\partial F}{\partial u_k}(\bar{0},\bar{0})\Big)\,\frac{\partial u_k}{\partial v_\ell}(\bar{0}) . \end{aligned} \tag{4.9}$$

Plainly (4.7) follows from (4.9).

The next question we consider concerns identifying a criterion for a function f to have a versal unfolding. We will observe that the existence of a versal unfolding of f is equivalent to an algebraic condition which is mediated by the transversality condition formulated above.

We now state without proof the following fundamental theorem, and then apply the notion of finite determinancy to show the converse of Theorem 4.7 .

Theorem 4.8. Let k be a non-negative integer. If $f \in m(n)$ is k-determined, then any two k-transversal unfoldings of f of the same codimension are equivalent.

There are five stages to the complete proof of this basic theorem. R. Thom in his I.H.E.S. notes has given a sketch of the basic idea behind the proof, but Wasserman [82] has given a completely rigorous proof.

Theorem 4.9. Let $f \in m(n)^2$ be k-determined. Then an unfolding (F,r) of f is versal if and only if it is k-transversal.

Proof: "Only if" is clear from Theorem 4.7 . Now let (G,s) be any unfolding of f. We must show that (G,s) is associated with (F,r). It is easy to see that, in the category of unfoldings of any germ f, there are always morphisms

$$(G,s) \to (G,s) + (F,r)$$

$$(F,r) + \text{constant} \to (F,r) .$$

Furthermore $(F,r) + (G,s)$ is obviously a k-transversal unfolding of f, since (F,r) is. For the same reason, $(F,r) + \text{constant}$ is also k-transversal. Then, by Theorem 4.8, we have an equivalence of unfoldings

$$(F,r) + \text{constant} \simeq (F,r) + (G,s)$$

where, on the left, we take the s codimensional constant unfolding. Thus we have

$$(F,r) \to (F,r) + \text{constant} \simeq (F,r) + (G,s) \to (G,s)$$

showing that (F,r) is versal.

Corollary 4.10. If for every k, the unfolding (F,r) of $f \in m(n)^2$ is k-transversal, then

(a) F is versal

(b) f is finitely determined.

Proof: By Theorem 4.9 all we have to show is that under the hypothesis, f is finitely determined. But if F is k-transversal, then F is right-left k-transversal, so by Theorem 4.1 we have for every non-negative integer k

$$\epsilon(n) = \langle\frac{\partial f}{\partial x}\rangle + f^*\epsilon(1) + W_F + m(n)^{k+1}$$

and hence $\dim_{\mathbb{R}} \epsilon(n)/\langle\frac{\partial f}{\partial x}\rangle + f^*\epsilon(1) + m(n)^{k+1} \leq \dim_{\mathbb{R}} W_F \leq r$, from which it is easy to see that f is finitely determined.

Theorem 4.11. (Existence of versal unfoldings) The germ $f \in m(n)^2$ has a versal unfolding if and only if it is finitely determined.

Proof: "If" is an immediate consequence of Theorem 4.9 and 4.6. "Only if" may be concluded from Theorem 4.7 and Corollary 4.10 in an obvious way.

Our algebraic criterion for an unfolding F of $f \in m(n)^2$ to be versal appears next and as one should suspect resembles a transversality condition previously developed.

Theorem 4.12. Let $f \in m(n)^2$ and let (F,r) be an unfolding of f. Then F is versal if and only if $\epsilon(n) = \langle\frac{\partial f}{\partial x}\rangle + V_F$.

The proof of this theorem is obvious from our previous theorems. We leave it to the reader.

As a result of this theorem, we can also calculate the minimal number of parameters that a universal unfolding of a finitely determined germ must have.

Recall that if $f \in m(n)^2$ is finitely determined, then

$$\text{cod } f = \dim_{\mathbb{R}} m(n)/\langle \frac{\partial f}{\partial x} \rangle_{m(n)} .$$

One may observe that each of these numbers are orbit invariants. That is, if f and g are in the same equivalent class, then $\text{cod } f = \text{cod } g$. As we indicated before, the terminology of codimension has indeed a geometrical justification, namely, if k is sufficiently large, then $\text{cod}(zL^k(n))$ ($\text{cod } L^k(1)zL^k(n)$) in $J^k(n,1)$ is equal to $n + r\text{-cod } f$ ($n + r\ell$ - cod f resepctively). Also, observe that $\dim_{\mathbb{R}} \varepsilon(n)/\langle \frac{\partial f}{\partial x} \rangle = (r\text{-cod } f) + 1$, since $\varepsilon(n) = \langle 1 \rangle_{\mathbb{R}} \oplus m(n)$

$$\varepsilon(n)/\langle \frac{\partial f}{\partial x} \rangle = m(n)/\langle \frac{\partial f}{\partial x} \rangle \oplus (\langle 1 \rangle_{\mathbb{R}} + \langle \frac{\partial f}{\partial x} \rangle)/\langle \frac{\partial f}{\partial x} \rangle .$$

One would then suspect that we have:

Theorem 4.13. For $f \in m(n)^2$ finitely determined, the minimal number of parameters appearing in a universal unfolding, i.e. the minimal codimension of an unfolding is the codimension of f.

Moreover, we also know how to construct universal unfoldings of f. The idea is the same as the proof in Theorem 4.6; we choose $b_1, b_2, \ldots, b_r \in m(n)$ as that their cosets in $m(n)/\langle \frac{\partial f}{\partial x} \rangle$ form a basis for this $\mathbb{R}$-vector space, and then $F(x, u_1, \ldots, u_r) = f(x) + \sum_{i=1}^{r} u_i b_i(x)$ is a universal unfolding of f.

This statement, and Theorem 4.3 would follow from a positive answer to the following question. Let k be a non-negative integer and let $f \in m(n)^2$ be k-determined. If we have two unfoldings (F,r) and (G,r) of f and F, G are k-transversal, is F associated with G? And if so, is G also associated with F? That is: can we induce F from G and vice versa with morphisms composing as the identity. Furthermore, if we can, then for a minimal dimension of unfolding (equal to $\dim_{\mathbb{R}} \varepsilon(n)/\langle \frac{\partial f}{\partial x} \rangle$), we can speak of a universal unfolding of f (the positive answer to this question is of course Theorem 4.8). If we could not, we would be content with the existence of versal unfoldings of f a notion which has also been found to be of value particularly in the works of

V. I. Arnold and G. N. Tyurina [2, 78].

Finally, we conclude this section with an important theorem in the development of singularity theory, namely, the Malgrange Preparation Theorem. We state the theorems below in Mather's formulation [41], which is a slight generalization of Malgrange's form.

Theorem 4.14. Let $f \in \varepsilon(n,p)$ with $f(0) = 0$. Let A be a finitely generated $\varepsilon(n)$-module and suppose $\dim_{\mathbb{R}} A/f^{*}m(p)A < \infty$. Then A is finitely generated as an $\varepsilon(p)$ module via f^{*}.

We will not prove this theorem in full generality; however, we will prove a special case of this theorem. The reason for doing so is that we believe the proof of this special case is straightforward and can also serve the purpose of illustrating the spirit of the general theorem. For a proof of Theorem 4.14, see e.g. [10, 20], or [82] or the articles of Wall, Nirenberg, Łojasiewicz, Mather, and Glaeser in [16, 39, 41, 57, 81].

Before we prove the special case of Malgrange Preparation Theorem, we need Mather's Division Theorem.

Theorem 4.15. Let F be a smooth real-valued function defined on a neighborhood of $\overline{0} \in \mathbb{R} \times \mathbb{R}^n$ such that $F(t,\overline{0}) = g(t)t^k$ where $g(0) \neq 0$ and g is smooth in a neighborhood of $0 \in \mathbb{R}$. Then given any smooth real-valued function G defined on a neighborhood of $\overline{0} \in \mathbb{R} \times \mathbb{R}^n$, there exists smooth functions q and r such that

1. $G = qF + r$ on a neighborhood of $\overline{0}$ in $\mathbb{R} \times \mathbb{R}^n$ and

2. $r(t,x) = \sum_{i=0}^{k-1} r_i(x)t^i$ for $(t,x) \in \mathbb{R} \times \mathbb{R}^n$ near 0.

For a proof, see [41].

Remark. This is only an existence-type theorem. If the functions F and G are analytic, then q and r are uniquely determined, which is the Weierstrass Preparation Theorem. However, with regard to the smooth category, the uniqueness no longer holds.

Example 4.6. $F(t,x) = t^2 - x$

$G(t,x) \equiv 0$.

Then $q = r = 0$ is one solution and the following is another solution:

$$r(t,x) = \begin{cases} e^{-1/x^2} & x \leq 0 \\ 0 & x \geq 0 \end{cases}$$

$$q(t,x) = -\frac{r(t,x)}{F(t,x)} .$$

Theorem 4.16. (Special case of the Malgrange Preparation Theorem)
Suppose that

1. M is a finitely generated $\varepsilon(n+1)$-module

2. $M/m(n)M$ is a finite dimensional real vector space.

Then M is a finitely generated $\varepsilon(n)$-module.

Before we give the proof, let us look at the following situation which will lead the reader to understand the significance of the theorem.

Let $M = \varepsilon(n+1)$/ideal generated by t^k, i.e. x is free and t is subject to $t^k \equiv 0$. M is certainly a finitely generated $\varepsilon(n+1)$-module. $M/\varepsilon(n)M$ is a real vector space generated by $1, t, \ldots, t^{k-1}$. Then this theorem concludes for us that M is a finitely generated module over $\varepsilon(n)$ with the set of generators $\{1, t, \ldots, t^{k-1}\}$.

Proof of Theorem 4.16: Let

$$M/m(n)M = \{[\sum_{i=1}^{k} r_i [v_i]] \mid r_i \in \mathbb{R}, \{[v_i]\} \text{ is the finite set of generators of } M/m(n)M\} .$$

Let $v_1, \ldots, v_k$ be representatives of $[v_1], \ldots, [v_k]$. Add $v_{k+1}, \ldots, v_\ell \in m(n)M$ such that $\{v_1, \ldots, v_k, v_{k+1}, \ldots, v_\ell\}$ form a set of generators of M over

$\epsilon(n+1)$. Let $t \in \mathbb{R}$, $tv_1 \in M$. Thus

$$tv_1 = \alpha_{11}v_1 + \dots + \alpha_{1k}v_k + \sum_{i=1}^{\ell} \beta_{1i}(x)v_i ,$$

where $\beta_{1i} \in m(n)$. This means that

$$[tv_1] = \alpha_{11}[v_1] + \dots + \alpha_{1k}[v_k] .$$

Then we have

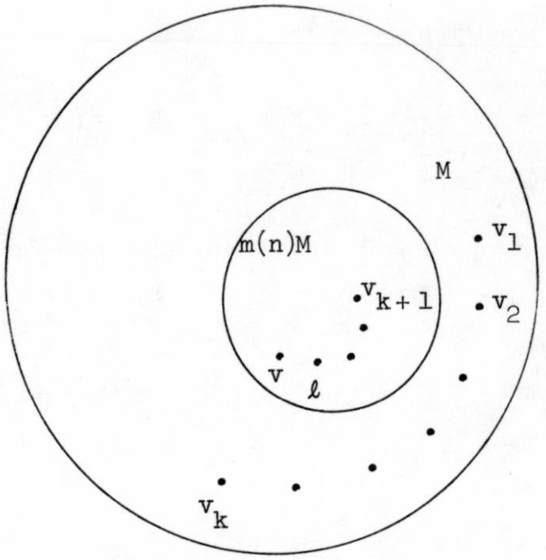

$$\begin{cases} (t-\alpha_{11}-\beta_{11})v_1 - (\alpha_{12}+\beta_{12})v_2 - \dots - (\alpha_{1k}+\beta_{1k})v_k - \beta_{1,k+1}v_{k+1} - \dots - \beta_{1\ell}v_\ell = 0 \\ \text{-------------------------------------} \\ \text{-------------------------------------} \\ (-\alpha_{\ell 1}-\beta_{\ell 1})v_1 - \dots\dots\dots\dots + (t-\beta_{\ell\ell})v_\ell = 0 \end{cases}$$

Let D be the coefficient matrix of this system of equations. Then

$$|D| = \det D = t^\ell + \dots$$

We now claim that $\{tv_1, \dots, tv_k, t^2v_1, \dots, t^2v_k, \dots, t^{\ell-1}v_1, \dots, t^{\ell-1}v_k\} = \mathfrak{D}$ is a set of $\epsilon(n)$-generator for M. For any $v \in M$, apply Theorem 4.15, we have

$$\begin{aligned} v &= f_1(x,t)v_1 + \dots + f_\ell(x,t)v_\ell \\ &= (q_1|D| + r_1)v_1 + \dots + (q_\ell|D| + r_\ell)v_\ell \\ &= r_1v_1 + \dots + r_\ell v_\ell, \text{ where } r_i = \sum_{i=0}^{\ell-1} r_{ij}(x)t^j , \\ &= \text{a combination of elements of } \mathfrak{D} \text{ with coefficients in} \\ &\quad \epsilon(n) . \end{aligned}$$

<u>Remark</u>. Certainly $|D|v_i = 0$ for each $i = 1, \dots, \ell$. This is clear by looking at the 2×2 situation, where

$$\begin{cases} ax + by = 0 \\ cx + dy = 0 \end{cases}$$

if and only if

$$\begin{vmatrix} a & b \\ c & d \end{vmatrix} x = \begin{vmatrix} a & b \\ c & d \end{vmatrix} y = 0 .$$

5. Thom's Classification Theorem

In this section, the following idea will be formalized: Given a "canonical unfolding" (G,r) (in a sense to be made precise later), and an arbitrary unfolding (F,k), $k \geq r$, of the same germ $f \in m(n)^3$, we should like to express the fact that the maxima and minima of F can be essentially gotten from the maxima and minima of G (as functions on $\mathbb{R}^n$ of course), that is maxima and minima of $F_{\bar{u}}|_{\mathbb{R}^n}$ for every $\bar{u} \in \mathbb{R}^k$ can be essentially gotten from the maxima and minima of $G_{\bar{v}}|_{\mathbb{R}^n}$ for each $\bar{v} \in \mathbb{R}^r$. In this way if F is a universal unfolding of f, then not only will it contain up to reparametrizations all unfoldings of f, but also will contain the entire configuration of extrema for unfoldings of f. Let us define the singular locus of the germ F at $\bar{0}$ to be the subset of $\mathbb{R}^{n+k}$ consisting of all singular points of the function $F|_{\mathbb{R}^n \times \{\bar{u}\}}$ (we do not distinguish the germ and its function representatives here) for $\bar{u}$ and $\bar{0}$ in $\mathbb{R}^k$. Clearly this definition is independent of the choice of the representatives of the germ F. This is why we make no distinction between the germ and its representatives. It is important to point out that the singular loci of unfoldings are of primary interest to the theory of singularities, not the unfoldings themselves. It is also obvious that the singular locus of F does not essentially change if we add a non-degenerate quadratic form $-y_1^2 - \dots - y_\mu^2 + y_{\mu+1}^2 + \dots + y_q^2$ to an unfolding F nor even if we take a constant unfolding of F. This leads us to the following definitions:

Definition 5.1. Let (G,r) be an unfolding of $g \in m(n)$. Let $(F,r+s)$

be an unfolding of $f \in m(n+q)$ and μ be an integer $0 \leq \mu \leq q$. F reduces to G with index μ if F is equivalent as an $r+s$ unfolding to the unfolding $(\hat{G}, n+q+r+s)$ where

$$\hat{G}(x,y,u,v) = G(x,u) - y_1^2 - \dots - y_\mu^2 + y_{\mu+1}^2 + \dots + y_q^2 \tag{5.1}$$

for $x \in \mathbb{R}^n$, $y \in \mathbb{R}^q$, $u \in \mathbb{R}^r$, $v \in \mathbb{R}^s$.

If F reduces to G with index μ, we call G a reduction of F with index μ.

Definition 5.2. F reduces to G properly if $q+s$ is positive in Definition 5.1, and G will be called a proper reduction of F. If an unfolding F has no proper reduction, then F will be said to be irreducible.

Remark. In case f and g are singular, and an unfolding $(F, r+s)$ of $f \in m(n+q)$ reduces to an unfolding (G,r) of $g \in m(n)$, then there exists an isomorphism (Φ, ψ, α), where $\Phi \in L(n+q+r+s)$, $\psi \in L(r+s)$ such that

$$F = \hat{G}\,\Phi + \alpha\pi_{r+s}, \tag{5.2}$$

where $\hat{G}$ is of the form (5.1). Thus μ is uniquely determined since μ, in this case, equals the index of the Hessian form of f minus the index of the Hessian form of g. (This last statement will be clarified from the Splitting Lemma.)

Corresponding to unfoldings, we have also a notion of reduction for germs:

Definition 5.3. Let $g \in m(n)$, $f \in m(n+q)$, μ be an integer $0 \leq \mu \leq q$. f reduces to g with index μ if f is equivalent to the germ $\hat{g}$ where $\hat{g}$ is defined by

$$\hat{g}(x,y) = g(x) - y_1^2 - \dots - y_\mu^2 + y_{\mu+1}^2 + \dots + y_q^2 \tag{5.3}$$

with $x \in \mathbb{R}^n$, $y \in \mathbb{R}^q$.

We call g a <u>reduction</u> of f with index μ if f reduces to g with index μ. Since index μ will not play an important role in the discussion, we say g is a <u>reduction</u> of f if for some μ, $0 \leq \mu \leq q$, f reduces to g with index μ.

<u>Definition 5.4</u>. f <u>properly reduces</u> to g if $q > 0$, in which case we say g is a <u>proper reduction</u> of f. f is <u>irreducible</u> if f has no proper reduction.

Our final notion prior to stating Thom's Classification Theorem is the correct notion of vacuity vis-a-vis reduction of germs of their unfoldings:

<u>Definition 5.5</u>. If (G,r) is an unfolding of $g \in m(n)$ and is an integer $0 \leq \nu \leq n$, G has a <u>simple singularity</u> with index ν at 0 if g is equivalent to $Q_\nu \in m(n)$ where

$$Q_\nu(x) = -x_1^2 - \dots - x_\nu^2 + x_{\nu+1}^2 + \dots + x_n^2 \qquad (5.4)$$

<u>Definition 5.6</u>. G has a <u>simple minimum</u> if $\nu = 0$ and G has a <u>simple maximum</u> if $\nu = n$.

Why is this a "vacuous" notion? g equivalent to Q_ν means $\langle \frac{\partial g}{\partial x} \rangle = \langle \frac{\partial Q_\nu}{\partial x} \rangle = \langle x_1, \dots, x_n \rangle = m(n)$. Hence, g is a universal unfolding of itself with codimension zero. Thus, G is isomorphic to a constant unfolding of g and hence, G reduces to the trivial unfolding $0 \in m(0)$. That is, G is equivalent to $0 + \sum_{i=1}^{q-\mu} y_{i+\mu}^2 - \sum_{i=1}^{\mu} y_i^2$, which will produce no catastrophe. Hence we must exclude this case when we state Thom's theorem.

Moreover if f has a local minimum (maximum) at 0, the index of the critical point must clearly at 0 (respectively n). Local minimality is what we are interested in. This can be seen easily from the following naive explanation. Consider a ball in a bowl (see Figure 5.1), the ball will remain at the bottom

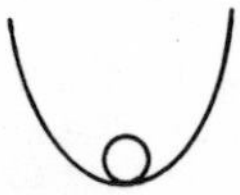

Figure 5.1.

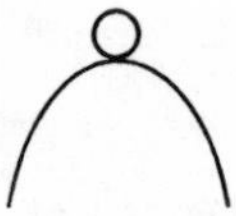

Figure 5.2.

(local minimum) of the bowl where it is in an equilibrium position and hence stable. If we flip over the bowl and place the ball at the top (local maximum) of the bowl, a slight perturbation of the ball will make it roll down the bowl, thus this situation is unstable. Let us look at the example $f(x) = x^4$ again, 0 is a local minimum of f, however 0 is not necessarily a local minimum of $F(x,u,v) = x^4 + ux^2 + vx$ although F has a <u>local minimum near</u> 0, this leads us to the following definition:

<u>Definition 5.7</u>. Let (F,r) be an unfolding of $f \in m(n)$. F has <u>local minima near</u> $\overline{0} \in \mathbb{R}^{n+r}$ if for every neighborhood W of $\overline{0}$ in $\mathbb{R}^{n+r}$ there is a point $(\overline{x},\overline{u}) \in W$ so that $F|_{(\mathbb{R}^n \times \{\overline{0}\}) \cap W}$ has a local minimum at $(\overline{x},\overline{u})$.

It is important to point out that if F has a local minimum at $\overline{0}$, then F has local minima near $\overline{0}$. However F can have local minima near $\overline{0}$ even if f has no minimum at $\overline{0}$. The obvious example for this case is $f(x) = x^3$ and $F(x,u) = x^3 + ux$.

<u>Theorem 5.1</u>. (Thom's Classification Theorem) Let $f \in m(n)^2$ be finitely determined, (F,r) a stable unfolding of f and has local minima near $\overline{0}$, and $r \leq 4$. Then either F has a simple minimum at $\overline{0}$, or F reduces with index 0 to one of the following seven irreducible (canonical) unfoldings G_i of germs g_i:

Name	g_i	G_i	Unfolding codimension
fold	$g_1(x) = x^3$	$G_1(x,u) = x^3 + ux$	1
cusp	$g_2(x) = x^4$	$G_2(x,u,v) = x^4 + ux^2 + vx$	2
swallowtail	$g_3(x) = x^5$	$G_3(x,u,v,w) =$ $x^5 + ux^3 + vx^2 + wx$	3
butterfly	$g_4(x) = x^6$	$G_4(x,u,v,w,t) =$ $x^6 + ux^4 + vx^3 + wx^2 + tx$	4
hyperbolic umbilic (wave crest)	$g_5(x,y) = x^3 + y^3$	$G_5(x,y,u,v,w) =$ $x^3 + y^3 + uxy + vx + wy$	3
elliptic umbilic (hair)	$g_6(x,y) = x^3 - xy^2$	$G_6(x,y,u,v,w) =$ $x^3 - xy^2 + u(x^2 + y^2) + vx + wy$	3
parabolic umbilic (mushroom)	$g_7(x,y) = x^2y + y^4$	$G_7(x,y,u,v,w,t) =$ $x^2y + y^4 + ux^2 + vy^2 + wx + ty$	4

Table 5.1.

Remark. If one is interested in the $r\ell$-category of unfoldings, reduction on both sides can be defined in a similar way as Definition 5.1. We only have to replace "equivalence" in Definition 5.1 by $r\ell$-equivalence. Let us observe the formalized conclusion of the two-sided case with $\mu = 0$. Then there exists an $r\ell$-isomorphism (Φ, ψ, λ), where $\Phi \in \varepsilon(n + q + r + s)$, so that $F(\bar{x},\bar{y},\bar{u},\bar{v}) = \lambda(\hat{G}\Phi, \pi_{r+s})(\bar{x},\bar{y},\bar{u},\bar{v}) = \lambda[\hat{G}(\varphi(\bar{x},\bar{y},\bar{u},\bar{v}), \psi(\bar{u},\bar{v})),\bar{u},\bar{v}]$ where $\varphi = (\varphi_1,\ldots,\varphi_n,\varphi_{n+1},\ldots,\varphi_{n+q}) = \pi_{n+q}\Phi$. By chain rule, it is not hard to find that

$$\frac{\partial F}{\partial x_i}(\bar{x},\bar{y},\bar{u},\bar{v}) = \frac{\partial\lambda}{\partial t}(\hat{G}\Phi, \pi_{r+s})(\bar{x},\bar{y},\bar{u},\bar{v}) \cdot$$

$$\left(\sum_{j=1}^{n} \frac{\partial G}{\partial x_j}(\bar{x},\bar{u}) \frac{\partial \varphi_j}{\partial x_i}(\bar{x},\bar{y},\bar{u},\bar{v}) + \sum_{\alpha=1}^{q} 2y_\alpha \frac{\partial \varphi_{n+\alpha}}{\partial x_i}(\bar{x},\bar{y},\bar{u},\bar{v})\right),$$

$$\frac{\partial F}{\partial y_k}(\bar{x},\bar{y},\bar{u},\bar{v}) = \frac{\partial \lambda}{\partial t}(\hat{G}\,\Phi\,,\pi_{r+s})(\bar{x},\bar{y},\bar{u},\bar{v}) \cdot$$

$$\left(\sum_{j=1}^{n} \frac{\partial G}{\partial x_j}(\bar{x},\bar{u}) \frac{\partial \varphi_j}{\partial y_k}(\bar{x},\bar{y},\bar{u},\bar{v}) + \sum_{\alpha=1}^{q} 2y_\alpha \frac{\partial \varphi_{n+\alpha}}{\partial y_k}(\bar{x},\bar{y},\bar{u},\bar{v}) \right),$$

for $1 \leq k \leq q$, $1 \leq i \leq n$.

To get the singular locus of F at $\bar{u} \in \mathbb{R}^r$, we suspend* the singular locus of $G|_{\mathbb{R}^n \times \{\bar{u}\}}$ with respect to $\{\bar{0}\} \times \mathbb{R}^s$ for $\bar{0}$ in $\mathbb{R}^q$. In order to keep the singular loci of F at $(\bar{u},\bar{v})$ and G at $\bar{u}$ being the same, we need $\frac{\partial \lambda}{\partial t} \neq 0$. Thus, we define that two $(r+s)$-codimensional unfoldings are <u>oriented equivalent</u> if they are $r\ell$-equivalent and $\frac{\partial \lambda}{\partial t} > 0$. Similar to Definition 5.1, we define that <u>F reduces orientedly to G with index μ</u>, $0 \leq \mu \leq q$, if we replace "equivalence" in Definition 5.1 by oriented equivalence. Thus, in this category, we say F reduces to G with index μ means that F reduces orientedly with index μ to G or $-G$. By means of oriented reduction one obtains another classification of unfoldings (in $r\ell$-category). As a matter of fact, unless $i = 2, 4$, or G_i and $-G_i$ are oriented equivalent. Therefore we let $g_8(x) = -x^4$, $G_8(x,u,v) = -x^4 + ux^2 + vx$; $g_9(x) = -x^6$, $G_9(x,u,v,w,t) = -x^6 + ux^4 + vx^3 + wx^2 + tx$ and $g_{10}(x,y) = x^2y - y^4$, $G_{10}(x,y,u,v,w,t) = x^2y - y^4 + ux^2 + vy^2 + wx + ty$. Then we have

<u>Theorem 5.2</u>. Let $f \in m(n)^2$ be finitely determined (F,r) a stable unfolding (in this case, it is a right-left universal unfolding) of f and has local minima near $\bar{0}$, and $r \leq 4$. Then either F has a simple minimum at $\bar{0}$, or F reduces orientedly with index 0 to one of the ten irreducible canonical unfoldings G_i of germs g_i, $1 \leq i \leq 10$. (They have been listed in Theorem 5.1 and the above mentioned G_8, G_9, and G_{10}.)

In other words, under oriented reduction, the list of canonical irreducible

---

*The suspension of a topological space M is defined to be the quotient space of $M \times [0,1]$ in which $M \times \{0\}$ is identified to be one point and $M \times \{1\}$ is identified to be another point.

unfolding consists of ten polynomials instead of seven. It is important to point out here the information about minima which plays a major rule in Thom's theory is preserved only by oriented reductions. However, Theorem 5.1 and 5.2 are corollaries to each other. It is obvious that if one merely is interested in reduction with index 0, then G_2 and G_8, as well as g_2 and g_8 are equivalent and also so are the pairs (G_4,G_9), (g_4,g_9), (G_7,G_{10}) and (g_7,g_{10}).

The proof of Theorem 5.2 will be given in Appendix II. The essence in the proof of the Classification Theorem involves correlating reductions of germs or singularities (elements of $m(n)^2$) to reduction of universal unfolding of such singularites. Thus, reduction of singularities will be discussed first.

Definition 5.8. For $f \in m(n)^2$, define the corank f to be $n - \text{rank}(\frac{\partial^2 f}{\partial x_i \partial x_j})(\overline{0})$.

This number thus measures the degree of degeneracy of f at the critical point $\overline{0} \in \mathbb{R}^n$.

Since the matrix $(\frac{\partial^2 f}{\partial x_i \partial x_j})(\overline{0})$ transforms under a change of coordinates $z = (z_1(x),\ldots,z_n(x))$ as

$$(\frac{\partial^2 f}{\partial z_i \partial z_j})(0) = {}^t(\frac{\partial z_i}{\partial x_j}(\overline{0}))(\frac{\partial^2 f}{\partial x_i \partial x_j}(\overline{0}))(\frac{\partial z_i}{\partial x_j}(\overline{0})), \tag{5.5}$$

we see that corank f is independent of the coordinate system used. Thus if $f \in m(n + q)$ reduces to $g \in m(n)$, then corank f = corank g and codim f = codim g. Furthermore, it is also easy to see that if $g \in m(n)^3$ is irreducible then the index of reduction by f depends only upon f, not upon g.

Theorem 5.3. (Residual Singularity) Let $f \in m(n)^2$ have corank p. Then there is $g \in m(p)^3$ to which f reduces.

Remark. Observe the improvement in Theorem 5.3 over a well-known corollary of the Morse lemma: if $f \in m(n)$ has corank p, then in a neighborhood of $\overline{0}$,

$$f(x_1,\ldots,x_n) = \sum_{i=1}^{n-p} \pm x_i^2 + \sum_{h,k=n-p+1}^{n} A_{h,k}(x_1,\ldots,x_n)x_h x_k \,. \qquad (5.6)$$

<u>Proof of Theorem 5.3</u>: In the usual cartesian coordinates, the matrix

$$A = \left(\frac{\partial^2 f}{\partial x_i \partial x_j}(\overline{0})\right) \qquad (5.7)$$

may be transformed into a matrix of the form

$$D = \begin{pmatrix} 0 & & & & & & & \\ & \ddots & & & & & \text{\Large 0} & \\ & & 0 & & & & & \\ & & & -2 & & & & \\ & & & & \ddots & & & \\ & & & & & -2 & & \\ & & & & & & 2 & \\ & \text{\Large 0} & & & & & & \ddots \\ & & & & & & & & 2 \end{pmatrix} \begin{matrix} \}\, p \\ \\ \}\, \mu \\ \\ \}\, n-p-\mu \,. \end{matrix}$$

That is, there is a non-singular matrix C such that $C^{-1}AC = D$. Then the matrix C considered as an element of $L^1(n)$ has a realization $\varphi_1 \in L(n)$ for which $f \circ \varphi_1$ has a Hessian matrix equal to D with respect to the usual cartesian coordinate system. Let μ = number of -2's on diagonal of D. Let $q = n - p$.

Define $Q \in m(q)^2$ by $Q(y_1,\ldots,y_q) = -y_1^2 - \ldots - y_\mu^2 + y_{\mu+1}^2 + \ldots + y_q^2$, where $(y_1,\ldots,y_q)$ is the usual cartesian coordinate on $\mathbb{R}^q$. Thus Q is 2-determined. It then follows that $g(y_1,\ldots,y_q) = f \circ \varphi_1(\overline{0},y_1,\ldots,y_q)$, where $\overline{0} \in \mathbb{R}^p$, has the same Hessian $(q \times q)$ matrix at $\overline{0}$ as Q. That is,

$$\begin{pmatrix} -2 & & & & \\ & \ddots & & & \\ & & -2 & & \\ & & & 2 & \\ & & & & \ddots \\ & & & & & 2 \end{pmatrix} \begin{matrix} \}\, \mu \\ \\ \\ \\ \\ \end{matrix} \,.$$

Since $\frac{\partial g}{\partial y_i}(\overline{0}) = 0$ for each i, the 2-jet at $\overline{0} \in \mathbb{R}^q$ of g and Q agree.

Since Q is 2-determined in $m(q)$, $q \sim Q$. Thus, there exists $\psi \in L(q)$ so that $g \circ \psi = Q$. That is, $g(\psi(y)) = f \circ \varphi_1(\overline{0}, \psi(y)) = Q(y)$. We can now introduce an unfolding in p parameters of Q by defining

$$\varphi_2(x,y) = \varphi_1(x, \psi(y)) . \tag{5.8}$$

Then $\varphi_2 \in L(n)$ and $f \circ \varphi_2(\overline{0},y) = f \circ \varphi_1(\overline{0}, \psi(y)) = g \circ \psi(y) = Q(y)$. Since Q is 2-determined, Q is a universal unfolding of itself. So, $f \circ \varphi_2$ is isomorphic to the p codimensional constant unfolding of Q, $\overline{Q}(x,y) = Q(y)$. Hence, there exists $\alpha \in m(p)$ and $\Phi \in L(n)$ so that $Q(y) = f \circ (\varphi_2 \circ \Phi)(x,y) + \alpha(x)$ (i.e. f reduces to α). In other words, $Q(y) - \alpha(x) = f \circ (\varphi_2 \circ \Phi)(x,y)$, where we now need to show $\alpha \in m(p)^3$.

Since f reduces to α, corank f = corank $\alpha = p$. From the reason $Q \in m(q)^2$, $f \in m(n)^2$ and the equation

$$\alpha(x) = Q(y) - f \circ (\varphi_2 \circ \Phi)(x,y) \tag{5.9}$$

we conclude that $\alpha \in m(p)^2$, and further the Hessian matrix of α must be zero, for otherwise the corank must be less than the dimension of the Euclidean space on which α is defined (i.e. corank $\alpha < p$). It follows that $\alpha \in m(p)^3$.

One observes from this proof that α is not constructible, and that to find such an α, other techniques must be resorted to.

When can we not reduce a germ $f \in m(n)^2$ further? Clearly, it now follows that f is irreducible if, and only if $f \in m(n)^3$.

As for the corresponding question asked of a universal unfolding F of a germ $f \in m(n)^2$, we have the following:

Corollary 5.4. A stable unfolding $(F, n + r)$ of $f \in m(n)^2$ is irreducible if, and only if $f \in m(n)^3$ and cod $f = r$. (Recall that the irreducibility of F means that F is not equivalent to $f(x) + Q(y)$ for $x \in \mathbb{R}^k$, $y \in \mathbb{R}^{n+r-k}$ for any $1 \leq k < n + r$.)

We now arrive at the basic classification theorem for germs of singularities underlying the Thom Classification Theorem.

<u>Theorem 5.5</u>. For any $f \in m(n)^2$ with $\operatorname{cod} f \leq 4$, f reduces to one of the following irreducible germs g_i:

Germ	Codimension	Corank
$g_0 \quad 0$	0	0
$g_1(x) = x^3$	1	1
$g_2(x) = x^4$	2	1
$g_3(x) = x^5$	3	1
$g_4(x) = x^6$	4	1
$g_5(x) = x^3 + y^3$	3	2
$g_6(x) = x^3 - xy^2$	3	2
$g_7(x) = x^2y + y^4$	4	2
$g_8(x) = -x^4$	3	1
$g_9(x) = -x^6$	4	1
$g_{10}(x) = x^2y - y^4$	4	2

Table 5.2.

The proof of Theorem 5.5 and hence Theorem 5.1 will be seen in Appendix II.

CHAPTER 4

CATASTROPHE THEORY

1. Introduction

In applied mathematics, engineering and physics, the subject of "shock waves" is plainly very important, since it appears in so many physical problems. Shock waves occur under a sudden change in some physical phenomenon (or physical system) from one state to another. Because of the discontinuous nature of a shock wave, it is a difficult phenomenon for physicists, engineers as well as mathematicians to comprehend. Recently, René Thom has classified the singularities for certain classes of functions, and this added enormously to our understanding of the qualitative aspect of discontinuity in natural processes. The main theoretical significance of Thom's classification is that it allows one to determine the stable equilibria of a gradient system subject to a small number of constraints, and to describe how these equilibria change as the constraints vary. This classification theorem is at the heart of catastrophe theory.

The usefulness of catastrophe theory obviously goes far beyond pure mathematics; in fact, Thom's methods have been applied to the analysis of problems in disciplines as diverse as biology, chemistry, engineering and sociology. The discussion in this chapter will be centered around catastrophe theory.

Many phenomena may be thought of as governed by a potential function of some form. The stable states of the system, i.e. those states which are actually observed to occur, may then be regarded as states for which some (potential) function is minimized. In the next section, examples will be provided to illustrate this point. If a function has multiple minima, then more than one stable state may be accessible to the system. Changing the control parameters in an experiment may alter the form of the governing potential in such a way as to change the positions, relative heights, or even total number of local minima. Thus, the stable states accessible to the system may change in a discontinuous way as the controls change smoothly. Observed discontinuous changes of state have been called "catastrophes." In section three, the notion of elementary catastrophe will be defined. Another form of Thom's Classification Theorem will be stated and further the seven types of elementary catastrophe will be listed. We will describe the delay rule and the Maxwell convention to illustrate how discontinuous change occurs in relation to local minima in section four. In this section we will not describe all the seven elementary catastrophe types, which we refer the readers to [10] or [99]. However we will describe the elementary catastrophes of corank one and corank two in detail with examples. Hopefully, from these illustrations the reader could build up a feeling about the physical meaning of the describing catastrophe types and the two rules mentioned above. In order to make the content in this chapter go smoothly, we will not get bogged down with the technical details. On the other hand, for the fully comprehension to Thom's theory, those details are essential. Thus, we will demonstrate the importance of the classification theorem together with Thom's three basic principles in morphogenesis in an Appendix I after Chapter 6.

2. Naive Discussion With Illustrative Examples

It is quite common in nature that the appearance (or existence) of a certain phenomenon may be interpreted as minimizing a potential function. As a matter of fact, the famous physicist Lagrange believed that every event in the world is of this kind! His belief was based on the fact that the motion corresponding to Newton's equation, or equivalently the Euler-Lagrange equation, is the result of minimizing the action integral $\int_{t_1}^{t_2} L dt$, where t is the time variable and $L(t)$ is the difference of kinetic energy and potential energy. Let us consider the following simple example.

Example 2.1. Catenary. Let a chain of length ℓ, of uniform density, be hung freely (i.e., under the influence of gravity only) at A, B, where A, B are located at the same level and a distance d apart

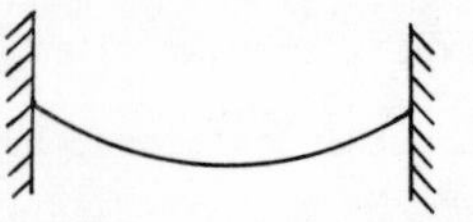

Figure 2.1.

(Figure 2.1). In this case, the kinetic energy is zero and the potential energy of a small segment of the chain is $\rho gyds$, where ρ is the density, g is the constance of gravity, y is the vertical coordinate and ds is the arc-length element $\sqrt{1 + (\frac{dy}{dx})^2}\, dx$. The action integral is $\int_0^d - \rho gy\sqrt{1 + (y')^2}\, dx$. This definite integral is to be minimized with the auxiliary condition $\int_0^d \sqrt{1 + (y')^2}\, dx = \ell$. In this case the Lagrangian multiplier method requires that we minimize the modified integral $-\rho g \int_0^d (y + \lambda)\sqrt{1 + (y')^2}\, dx$. Following the method of Euler-Lagrange, the problem of minimizing a definite integral which contains an unknown function and its derivative can be reduced to the elementary problem of minimizing a function of several variables. It is here that the famous Euler-Lagrange differential equation enters. The necessary and sufficient condition for the integral $\int_0^d F(x,y,y')dx$ to be stationary, with the boundary condition $y(0) = \alpha$, $y(d) = \beta$ (in our case α, β are zero) is that the differential equation of Euler-Lagrange $\frac{\partial F}{\partial y} - \frac{d}{dx}\frac{\partial F}{\partial y} = 0$ shall be satisfied. With this minimization process we will get the differential equation $\frac{y''}{\sqrt{1 + (y')^2}} = a$, where a is a constant. The solution curve is the well known catenary $y = \frac{1}{a}\cosh(ax)$, which in fact is the shape of the chain hanging over A, B under the influence of gravity. This fact indicates that the natural phenomenon -- the shape of a freely hanging uniform density chain -- may be explained by the minimization of a potential function.

Let us look at some more examples which can be understood just by common sense. However, we must realize that, in most situations, the precise form of the potential function is not known.

<u>Exampel 2.2</u>. Consider the path of light travelling form A to B (Figure 2.2), where A is in one medium (say air) and B is in another medium (say water). Let us imagine that there are infinitely many possible paths that

the light ray will travel from A to B. However, from common sense (or physics) we know that the path the light ray travels minimizes the time for the light to travel. Formally, we define the state space X to be the set of all paths from A to B. (X is an infinite dimensional manifold.) Define the potential function $V: X \to R$ to give the time for light to travel along the given path. The natural phenomenon is then explained by the fact that the light ray travels in the shortest time interval.

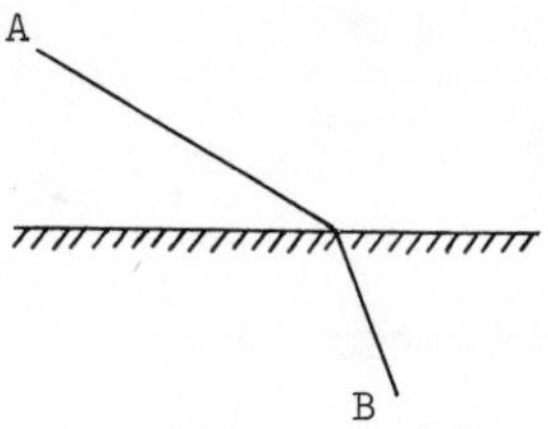

Figure 2.2.

Example 2.3. Let us immerse a ring (regular shape or not) in soapy water and pull it out slowly. We will get a surface of minimum surface area (plane or a bubble according to the shape of the ring). The surface under consideration can also be looked at as maximizing surface tension. In this case the state space is the set of all possible shapes (which is again an infinite dimensional manifold), the value of the potential function $V: X \to R$ is the surface area. In analogy with the other examples, the natural phenomenon is that in which the surface area has been minimized, or equivalently, the surface tension has been maximized.

Example 2.4. A principle in architecture is to minimize the cost of construction and at the same time to maximize the usefulness and the aesthetic quality of the structure. Similarly any business is trying to maximize net profit.

With these examples in mind we will clearly be interested in studying the locations of the extrema of potential functions V, i.e. those x^* where $V(x^*)$ is an extremum.

On the other hand, many phenomena in nature exhibit discontinuities. Those discontinuous phenomena, which recur if one repeats the experiment with the same initial data, are the most interesting. Let us illustrate some discontinuous phenomena with examples.

Example 2.5. Boiling water. In the case of water boiling in a pot, the discontinuity of this phenomenon consists in the fact that the density of the substance in the pot will suddenly change from one to zero at the time when the water boils. The qualitative behavior will not be different if we perturb the pressure in the pot a little; the discontinuity of the phenomenon (i.e. the sudden change of density) will always occur. This type of discontinuity is stable. (This is a particular case of a more detailed discussion in Example 4.2 .)

Example 2.6. Breaking of waves [102] . Waves of water will break when they reach the shore (Figure 2.3).

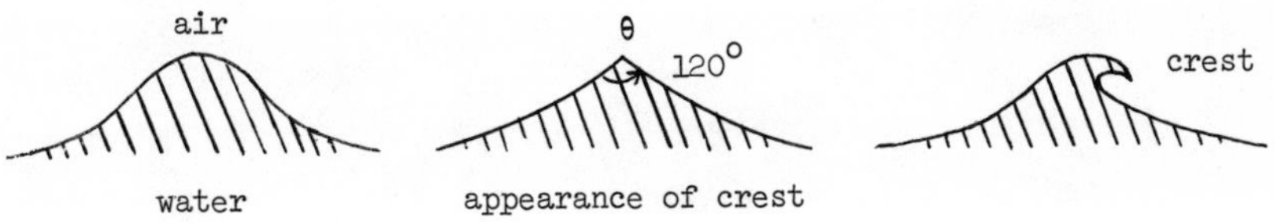

Figure 2.3.

Whatever the angle, θ , of break, there is a cusp immediately after. Furthermore, the angle θ has been shown by Stokes to be 120° , and shallow water theory shows the wave is symmetric before break -- both results have been confirmed by observation. The qualitative behavior of this phenomenon can be explained by combining classical hydrodynamics with Thom's catastrophe theory.

Example 2.7. A wind tunnel with a narrow neck. Let the wind be blowing in a wind tunnel with a narrow neck (Figure 2.4). The velocity at B is higher than that at A . Therefore the pressure on the wall of the tunnel at B is smaller than that at A . As v_0 increases, v_1 at B increases and the pressure at B decreases accordingly. As v_0 increases to a certain level, v_1 reaches v_1^*, supersonic velocity. At this point, the graph of the pressure against velocity forms a cusp. A film about this phenomenon indeed revealed the discontinuous behavior of the pressure diagram at B . Instead of a curve at the cusp point, we get a bunch of points where we cannot decide the graph of the shock wave. In fact, this kind of discontinuous phenomenon happens in many physical problems. In case the catastrophe set is simple enough (see next section), Thom's theorem is applicable. Through the classification theorem we can determine the types of the catastrophe sets.

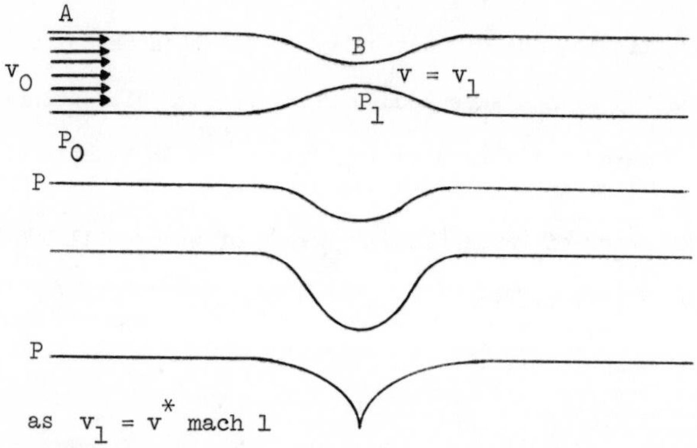

Figure 2.4.

3. The Elementary Catastrophes

The elementary catastrophes are certain singularities of smooth mappings of codimension less than or equal to four. They are important because they occur in our space-time in a structurally stable way.

Let $V: \mathbb{R}^n \times \mathbb{R}^r \to \mathbb{R}$ be a potential function, where $\mathbb{R}^r$ is called the parameter space (or control space) and $\mathbb{R}^n$ is called the state space. V can also be visualized as a smooth family V^u, indexed by $u \in \mathbb{R}^r$, of potential functions defined on $\mathbb{R}^n$. In Thom's theory of catastrophes, the mathematical model for a natural process is obtained in many cases by considering the set of minima of such a family V^u of potential functions. Locally, near a minimum, such a family of potential functions is of course just an unfolding of a singular germ. In case $r \leq 4$ and supposing the potential functions V under consideration are "structurally stable," Thom claims, in his classification theorem, that there are precisely seven essentially different (up to diffeomorphism) unfoldings which can be involved in the local description of a natural process.

Further, a catastrophe set is the set of points in the parameter space where the location of the minima of potential functions undergoes a sudden change (i.e., a discontinuity) as the parameters vary. The central theme of catastrophe theory is the study of the topological type, up to diffeomorphism, of these catastrophe sets. It turns out that the local study of these sets is equivalent to the local study of the unfoldings. As such, we may direct our attention to the unfoldings and formulate our conclusions with these objects as compared to their corresponding catastrophe ("bifurcation") sets. These seven types of catastrophe sets are called the elementary catastrophe. A further remark to be made here is that the condition that the potential functions under consideration are locally stable (as unfoldings) is a natural one, since in the physical world only structural stable phenomena are meaningful objects of scientific study.

The following is another form of Thom's Classification Theorem for universal unfoldings. The equivalence between this form and the one presented in Chapter 3 can be seen in Appendix I .

In the set

$$\mathfrak{F} = \{V: \mathbb{R}^{n+r} \to \mathbb{R},\ V \text{ smooth},\ r \leq 4\} \tag{3.1}$$

there is an open dense subset $\mathfrak{F}_0$ consisting of maps $V: \mathbb{R}^{n+r} \to \mathbb{R}$ such that, if $(p,q) \in \{(x,u) \in \mathbb{R}^{n+r} \mid x = (x_1,\ldots,x_n) \in \mathbb{R}^n,\ u \in \mathbb{R}^r$ and $\frac{\partial V}{\partial x_1}(x,u) = \frac{\partial V}{\partial x_2}(x,u) = \ldots = \frac{\partial V}{\partial x_n}(x,u) = 0\}$, then, in a neighborhood N of (p,q), $V|_N$ is a representative of a universal unfolding of the representative $\eta = V|_{N \cap (\mathbb{R}^n \times \{\bar{0}\})}$ of a germ of codimension $\delta \leq 4$. (However, V may not be of minimal unfolding codimension δ.) Further, the set

$$M_V = \{(x,u) \mid \frac{\partial V}{\partial x_1}(x,u) = \ldots = \frac{\partial V}{\partial x_n}(x,u) = 0\} \tag{3.2}$$

is an r-dimensional manifold for all V in this large set $\mathfrak{F}_0$. Thus, generically, a potential function $V: \mathbb{R}^{n+r} \to \mathbb{R}$, considered as a family of parametrized potential functions $V^u: \mathbb{R}^n \to \mathbb{R}$ for $u \in \mathbb{R}^r$ (we adopt from now on the notation using superscripts as parametrized potential functions parametrized by elements in $\mathbb{R}^r$ - the parameter space), possesses the property that the set of parametrized equilibrium positions forms, as a subset of $\mathbb{R}^{n+r}$, an r-dimensional manifold. The second basic conclusion emanating from this far reaching generalization of Whitney's theorem described in Chapter Two is the classification of germs (with less than or equal to two variables) of codimension ≤ 4. Finally if we let $\chi_V: M_V \to \mathbb{R}^r$ be the map induced by the projection $\mathbb{R}^{n+r} \to \mathbb{R}^r$, we call χ_V the <u>catastrophe map</u> of V and then any singularity χ_V is equivalent to one of the seven types of elementary catastrophe (in case $r \leq 4$). The clarification of the last statement and a more detailed discussion of Thom's theory of catastrophes will be found in Appendix I. The list of these seven elementary catastrophes is as follows:

Name	Germ	Universal Unfolding	Codimension
(1) The fold	x^3	$x^3 + ux$	1
(2) The cusp (Riemann-Hugoniot)	x^4	$x^4 + ux^2 + vx$	2
(3) The swallow tail	x^5	$x^5 + ux^3 + vx^2 + wx$	3
(4) The butterfly	x^6	$x^6 + ux^4 + vx^3 + wx^2 + tx$	4
(5) The hyperbolic umbilic	$x^3 + y^3$	$x^3 + y^3 + wxy + ux + vy$	3
(6) The elliptic umbilic	$x^3 + 3xy^2$	$x^3 + 3xy^2 + w(x^2 + y^2) + ux + vy$	3
(7) The parabolic umbilic	$x^2y + y^4$	$x^2y + y^4 + wx^2 + ty^2 + ux + vy$	4

The proof of this theorem can be found in Appendix II or in [10, 82, 99]; we will not furnish it in this chapter. However, it will be our goal in the rest of this chapter to study the interpretation of Thom's claim and to explain the meaning of the two simplest catastrophe types of the above list. The pictures of the other catastrophe types (under delay rule) can be found in either [60] or a recent published book [10]. For those who do not read Chapter 3, let us look at the above list again. It is natural to ask why the representation of the germs of elementary catastrophes only involve at most two variables. The answer to this question, by the following reduction lemma, is a key step in allowing us to reduce the elementary catastrophes to a finite number. Furthermore, the reduction lemma is the link between the global and local parts of the classification theorem; that is, between the genericity conditions of $\mathfrak{F}_0$ and the finite classification of the local representation of the members of $\mathfrak{F}_0$. We leave the detail explanation of this statement in Appendix I.

Lemma 3.1. (Reduction Lemma). Let $S = \{Z \in J^r(n,1) \mid j^{(1)}(z) = 0\}$. For any generic differentiable mapping

$$\sigma\colon\ \mathbb{R}^4 \to J^r(n,1)$$

with $\mathrm{Im}(\sigma) \subset S$. Then we can write $\bar{u} = (u,v,w,t)$ and

$$\sigma(\bar{u})(x) = \sum_{i=3}^{n} (\pm x_i^2) + \rho(\bar{u};\, x_1,x_2) \tag{3.3}$$

in a neighborhood of any given point $(\bar{u},0)$ in $\mathbb{R}^{4+n}$, with $x = (x_1,\ldots,x_u) \in \mathbb{R}^n$ and $\bar{0} \in \mathbb{R}^n$. Further for each $\bar{u} \in \mathbb{R}^4$, $\rho_{\bar{u}}\colon \mathbb{R}^2 \to \mathbb{R}^1$ defined as $\rho_{\bar{u}}(x_1,x_2) = \rho(\bar{u};\, x_1,x_2)$ is in $m(2)^3$.

Proof: For $z \in S$, write

$$z(x) = Q(x) + (\text{terms of degree } \geq 3) \tag{3.4}$$

where Q is a quadratic form. Let the rank of Q be $n-k$, so that the corank of Q is k; then by a suitable linear transformation of coordinate systems we can write

$$Q(x) = \sum_{i=k+1}^{n} (\pm x_i^2)\,. \tag{3.5}$$

In this new coordinate system,

$$z(x) = \sum_{i=k+1}^{n} (\pm x_i^2) + z_1(x)\,, \tag{3.6}$$

where $j^{(2)}(z_1) = 0$ and z_1 is a function of $x_1, x_2, \ldots, x_n$. Changing coordinates again [37, Theorem 4], $z(x)$ can be reduced to $\sum_{i=k+1}^{n} (\pm x_i^2)$ $\sum_{i=k+1}^{n} (\pm x_i^2) + z_2(x)$ where $z_2(x)$ does not involve any power of x_{k+1} of degree ≥ 2. Then, with one more change of coordinates $z_2(x)$ will not involve x_{k+1}. Similarly, $x_{k+2}, \ldots, x_n$ can also be absorbed into $\sum_{i=k+1}^{n} (\pm x_i^2)$ by means of a sequence of changes of coordinates. Thus $z(x)$ can be written as

$$z(x) = \sum_{i=k+1}^{n} (\pm x_i^2) + \rho(x_1, \ldots, x_k) \tag{3.7}$$

where $j^{(2)}(\rho) = 0$. We call ρ the <u>residual singularity</u> [74, Chapter 5].

Let L_k denote the submanifold of S consisting of all $z \in J^r(n,1)$ with corank $Q = k$, then $\operatorname{codim} L_k = \frac{k(k+1)}{2}$. This is because the condition for a critical point of $z(x_1, \ldots, x_n)$ to be of corank k, $k \leq n$, is that the coefficients of the terms $x_i x_j$ with $1 \leq i \leq j \leq k$ in the quadratic part of z should vanish. This imposes $\frac{1}{2}k(k+1)$ independent conditions. Thus $\operatorname{codim} L_k = \frac{k(k+1)}{2}$. In particular, if $k \geq 3$, then $\operatorname{codim} L_k \geq 6$ and if $k \leq 2$ $\operatorname{codim} L_k \leq 3$. Hence, by the Transversality Theorem, we conclude that for a generic differential mapping $\sigma: R^4 \to J^r(n,1)$, the image of σ will miss L_k for any $k \geq 3$. This shows that generically $\sigma(u,v,w,t)(x)$ can be written as $\sum_{i=3}^{n} \pm x_i^2 + \rho(x_1, x_2, u, v, w, t)$ in a neighborhood of any point in R^4.

4. Types of Elementary Catastrophes

We are now going to describe the topological structure of shock waves, in particular to describe the singularities of codimension less than or equal to two, with examples; and we will consider strata (of the function space) of conflict and bifurcation. These descriptions will be useful in practice. Usually when a shock wave occurs it is impossible to get a clear picture of what happens in the neighborhood around that of the shock wave. With the help of the list of seven catastrophe types, we can identify the particular type which occurs, and, thus, we can analyze the behavior near the singular point.

The stable local regimes at a point $u \in R^4$ are defined by the structurally stable attractors of $-\text{grad } V^u$, that is, by the minima of the potential function V^u. In general, there will be several such minima, of which only one can dominate. The question is how this one should be chosen from all the theoretically possible regimes. There are two simple conventions to be described in the following. Examples will follow which illustrate how and when to use them.

(1) Maxwell's Convention. This convention or rule is named after the 19th century Scottish physicist J. C. Maxwell, who formulated a similar rule in thermodynamics when the Van der Waals' equation [13, 14] was used to describe the catastrophic jump in volume (or density) as a liquid boils (see Example 4.2 of this chapter). The convention asserts that, when two or more stable attractors are in competition at a point u in the parameter space, the chosen state is the one of absolute minimum potential. Catastrophic conflict arises only when there are more than one absolute minimum for the potential function.

(2) Delay Rule. This convention or rule is that any one of the stable attractors is to be chosen at will, and then to be systematically (continuously) chosen (as u varies), until it disappears into a degenerate critical point. At this moment the minimum will make a sudden change. Thus the locus of parameter u such that the potential function V^u has degenerate critical points assume a role of importance.

Example 4.1. $V(x,u) = \frac{1}{3} x^3 + \epsilon x^4 + ux$ (4.1)

with $\epsilon > 0$ small.

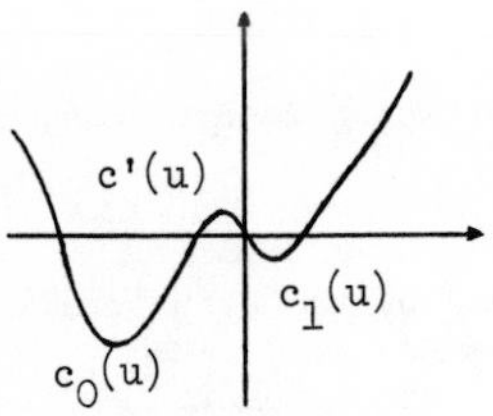

Figure 4.1.

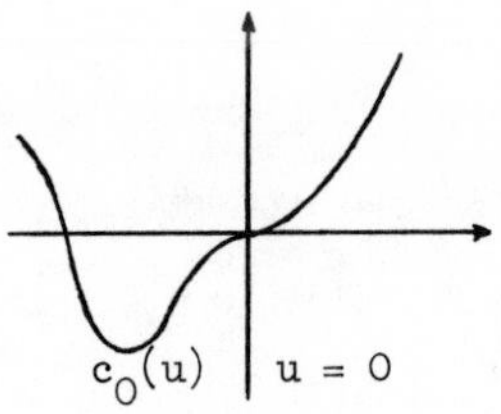

Figure 4.2.

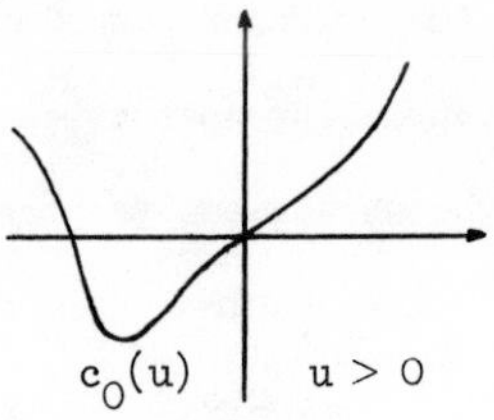

Figure 4.3.

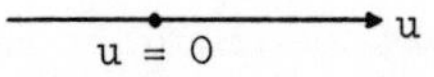

Figure 4.4.

This potential is the universal unfolding of the singularity $V = \frac{1}{3} x^3$. The compactifying term ϵx^4 has no effect on the minimum at the critical point $(0,0)$ when $u = 0$, if we only consider a small neighborhood of the origin. In fact, if without the term ϵx^4, the figure have been shown in Figure 1.2 of Chapter One. However, it produces a minimum point $c_0(u)$ at a rather large distance from the origin. This artificial term is just for our convenience to illustrate the shock wave phenomenon. Without this term, it happens sometime, when $u \geq 0$, that we do not have any minimum, which is not a desirable situation when discussing catastrophe types.

In Figure 4.1, noticing that $c'(u)$ is not a stable attractor, there are only two stable attractors, $c_0(u)$ and $c_1(u)$, that is, two local minima. Under Maxwell's convention, this potential will produce no catastrophic situation of "conflict," i.e., for no u is it true that $V^u(c_0(u)) = V^u(c_1(u))$ where $c_0(u)$ and $c_1(u)$ are the two (at most) local minima of V^u. Thus, Maxwell's convention is just imapplicable here. Under the delay rule, if we choose the stable attractor $c_0(u)$ at the beginning. There is no catastrophic situation either, as u varies, since for any u_0, $\lim_{u \to u_0} c_0(u) = c(u_0)$ is not a degenerate critical point. However, if we choose c_1 first, when $u < 0$, then as u reaches 0, the stable minimum $c_1(u)$ disappears into a degenerate critical point. At this point the minimum will jump to $c_0(0)$. This produces a shock wave, at $u = 0$, called the point shock wave (as in Figure 4.4) which is the simplest elementary catastrophe. This mathematical model fits nicely with the physical model of dropping a stone on the floor, although we admit that it is a trivial example. Even though we would like to ask beginners that what is the potential function in this case.

Example 4.2.

$$G(x,u,v) = \frac{1}{4}x^4 + u\frac{x^2}{2} + vx \tag{4.2}$$

This potential is the universal unfolding of the singularity $G = \frac{1}{4}x^4$. This is the widely publicized cusp catastrophe [56, 67, 101], which is the next simplest one. The catastrophe set of this potential function, under delay rule, is indeed a cusp. It can be derived as follows.

Consider

$$\frac{\partial G}{\partial x} = x^3 + ux + v \tag{4.3}$$

then the two-dimensional manifold is given by

$$M_G = \{(x,u,v) \mid x^3 + ux + v = 0\} \tag{4.4}$$

and the catastrophe map, $\chi_G : M_G \to \mathbb{R}^2$, is induced by the projection $\mathbb{R}^{1+2} \to \mathbb{R}^2$ onto the parameter space (see Figure 4.5).

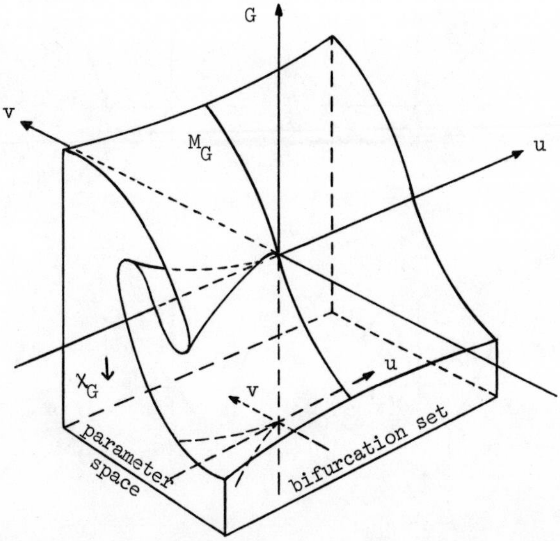

Figure 4.5.

The catastrophe set of G is the set of singular values of χ_G, and can be obtained by setting either $\frac{\partial G}{\partial x} = 0$ with $\frac{\partial^2 G}{\partial x^2} = 0$ or the discriminant of the equation $\frac{\partial G}{\partial x} = 0$ (i.e. eliminating x). In either case, we get a curve $\{(u,v) \mid 4u^3 + 27v^2 = 0\}$, which is a cusp in the (u,v) plane. For simplification we will use P or $P(u,v)$ as a point in the parameter space with coordinates u and v.

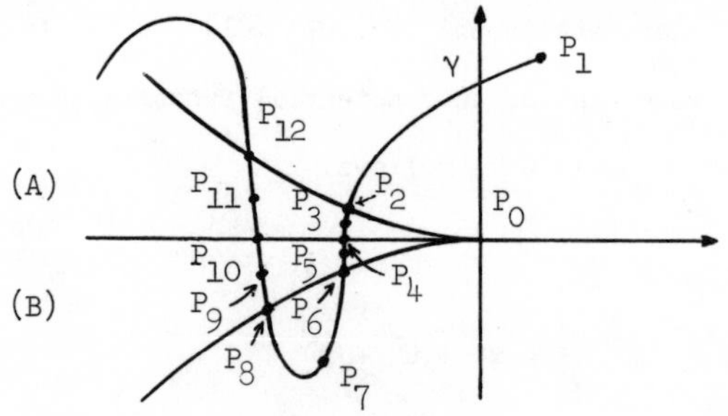

Figure 4.6.

Corresponding to each point in the parameter (u,v) space, there is a potential. Let $P_0 = P(0,0)$, $P_1 = P(u_1,v_1), \ldots, P_7 = P(u_7,v_7)$ be eight representative point in the $u - v$ plane (see Figure 4.6), then the corresponding potentials are as in Figures 4.7 - 4.14.

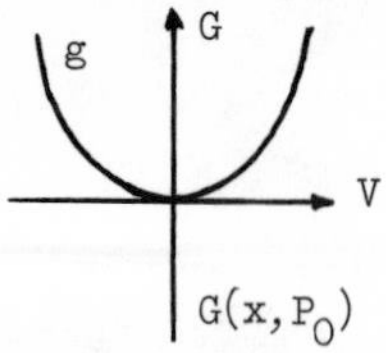

Figure 4.7.

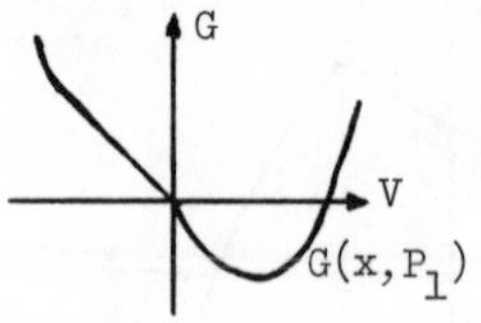

Figure 4.8.

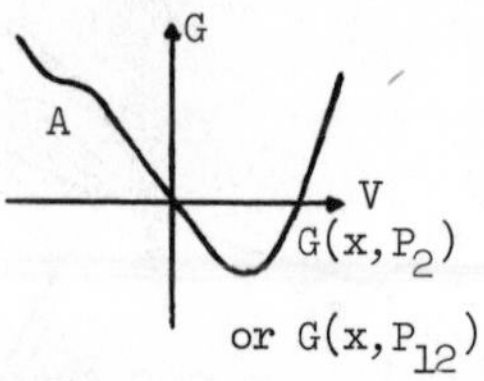

Figure 4.9.

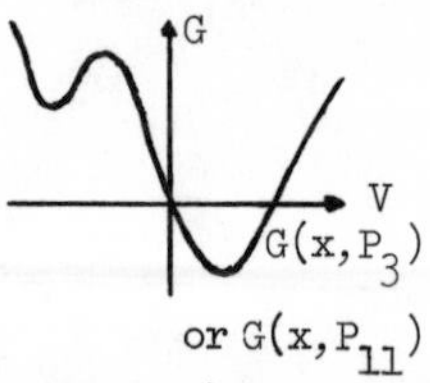

Figure 4.10.

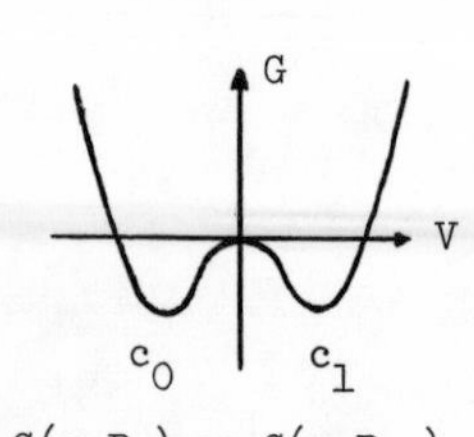

Figure 4.11.

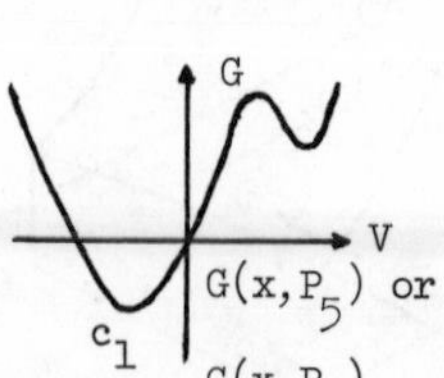

Figure 4.12.

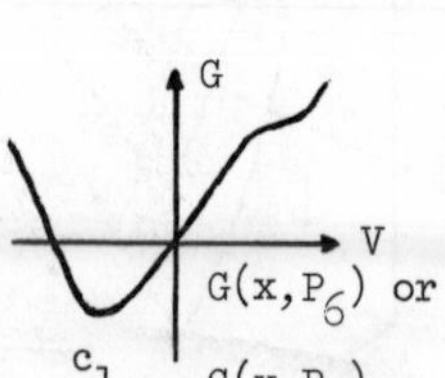

Figure 4.13.

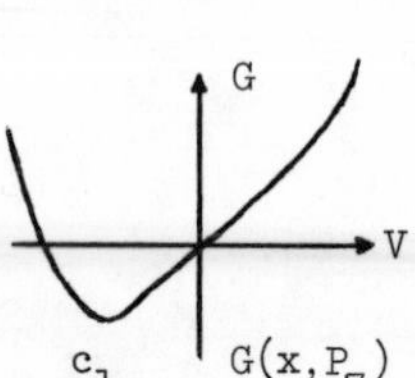

Figure 4.14.

Under delay rule the catastrophe set in this example is the cusp as in Figure 4.6. Let us explain this fact as follows. Starting from any point, say P_3 in the region A, in the parameter space, the corresponding potential is $V(x,P^3)$ as in Figure 4.10. In Figure 4.10, we could choose either one of the local minima $c_0(P_3)$, $c_1(P_3)$ in the beginning. Let us choose $c_0(P_3)$. As $P(u,v)$ varies in the parameter space, the potentials vary accordingly. As P approaches toward P_4 (see Figure 4.11), according to the delay rule the choice of the local minimum will still be $c_0(P_4)$. In other words, at P_4, there is no sudden change of the locations of local minima, thus P_4 is not a point in the catastrophe set. Only when P reaches P_6, a point on the lower branch of the cusp (see Figure 4.13), does the local minimum $c_0(P)$ disappear and the stable local regime at $c_1(P_6)$ will be chosen. In this way therefore, the shock wave phenomenon is formed, by letting P vary in the (u,v) plane. Let us move P along the curve γ in Figure 4.6. So long as P does not reach to the upper branch of the cusp, there is no catastrophic change. We explain this as follows. As $P \to P_8$, $c_1(P) \to c_1(P_8)$ which is still a stable minimum of G^{P_8}. Similarly, as $P \to P_{10}$, $c_1(P) \to c_1(P_{10})$, again a stable minimum. Therefore no catastrophic change in state occurs. But when $P \to P_{12}$ (or P_2) (see Figure 4.9), a point on the upper branch of the cusp, $c_1(P) \to c_1(P_{12})$, a degenerate critical point of $G^{P_{12}}$ or G^{P_2}. Hence there is a sudden change from $c_1(P_{12})$ to the stable regime $c_0(P_{12})$. Thus, the catastrophe set under the delay rule would be the cusp, which has also been named as the Riemann-Hugoniot* catastrophe set. One can get a feeling for this behavior if one imagines the negative u-axis to be the positive time axis, so $u = -t$. If v is some spatial variable, measuring distance of an object from a fixed origin then one might see a path like Figure 4.15. For the applications of this catastrophe type we refer the reader to [14, 19, 21, 22, 25, 26, 63, 68, 93, 95, 97, 100, 101, 102].

*Hugoniot studied the shock wave phenomena of gas dynamics as the formation of a shock wave when an acceterated piston travelled in a cylinder. The possibility of such a shock wave had been predicated theoretically by Riemann.

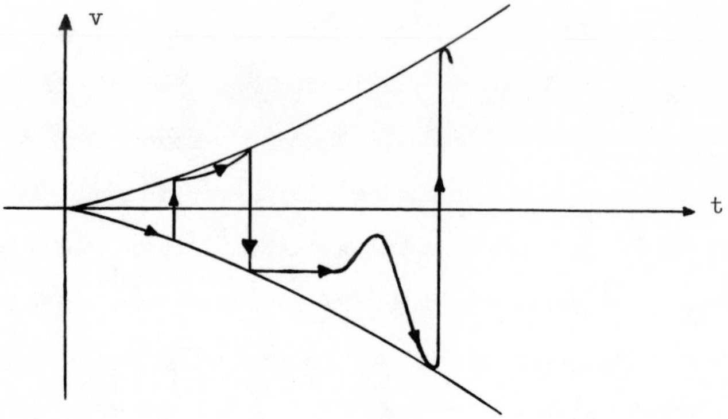

Figure 4.15.

Under Maxwell's convention the catastrophe set in this example (Example 4.2) is the negative u-axis (see Figure 4.6) since that is the only place the stable attractors are in competition (i.e. $G^{P_4}(x_1) = G^{P_4}(x_2)$) and thus the sudden change in locations of local minima takes place. The process of phase transition in the Van der Waals Model offers a good example of a process analyzable with Thom's formalism.

Consider a pure gas (not a mixture of various gases, although qualitatively our analysis would not be radically different if the sample were not pure) in a cylinder with movable piston. We place this cylinder into a container with insulating walls. Thus, the walls of the cylinder but not of the container allow for a heat flow. We further are able to exert a pressure on the piston and thus alter the internal pressure and volume of the gas. One uses this apparatus (see Figure 4.16) to derive the pressure (P) - volume (V) relationship of the gas during a phase change from gas to liquid.

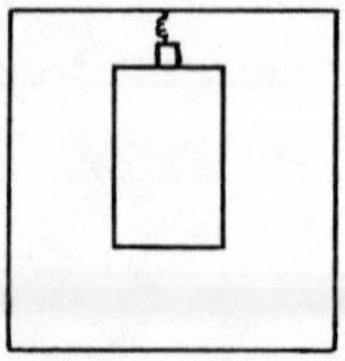

Figure 4.16.

If one is familiar with the classical attempts at describing mathematically the states of a real gas during such a transition, for example, Clausius' or

Van der Waals' equations [13] , one expects a P - V diagram to look like

Figure 4.17 at a constant temperature T below a certain critical temperature.

Here, the region marked α is never observed owing to the instability (physically

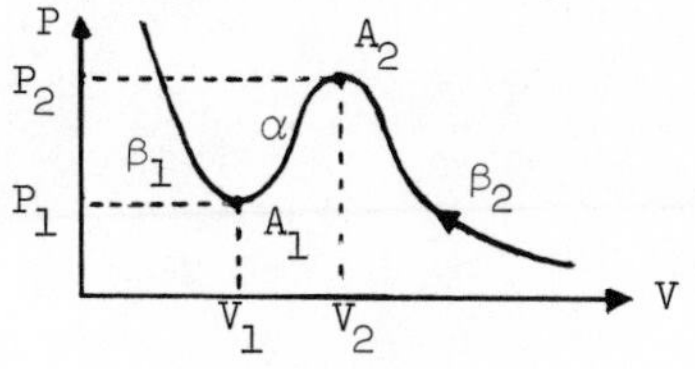

Figure 4.17.

$\frac{dV}{dP} > 0$ cannot happen) of these states. Moreover, there are two experimental

observations which serve as supporting material for Thom's underlying hypothesis.

If one starts at high volume low pressure, the gas phase, and begins to

exert external pressure on the piston (always maintaining a constant internal

temperature), thereby increasing the internal pressure of the gas, one will move

along the branch β_2 of Figure 4.17 . Assume one is reasonably careful

experimentally, insuring no impurities in the cylinder, isolation from external

vibrations etc., one observes that at a certain fixed pressure P_τ there will

ensue a continual phase change as the gas condenses into a liquid. Then, any

subsequent pressure increase merely decrease the volume of the liquid along the

branch β_1 . Thus, what one really observes is a graph like Figure 4.18, where

P_τ is in between P_1 and P_2 .

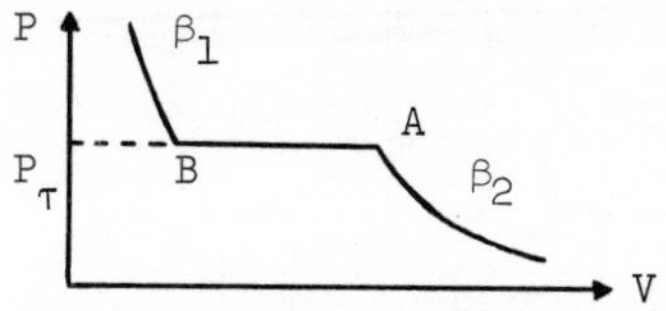

Figure 4.18.

The second observation one can make involves the superheating S_1 and

supercooling S_2 states of the liquid or gas. These are states corresponding

to the portions of Figure 4.17 labelled δ_1 and δ_2 respectively in Figure 4.19 .

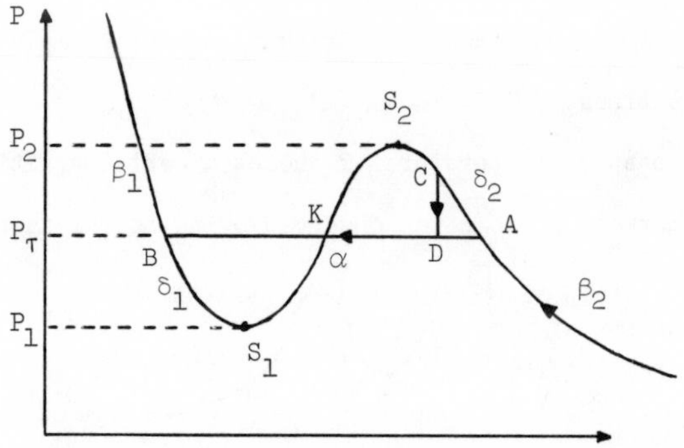

Figure 4.19.

By various subtle and difficult experimental devices, one is able to go from the branch β_2 up through any state on the arc δ_2. Similarly, in traversing a path from liquid to gas, one can go from the branch β_1 down to any point on the arc δ_1. However, if one is at a state C on the arc δ_2, any tiny perturbation (like vibration, etc.) the state C will drop down to D and then change phase at the pressure P_T, ending up at state B. Then, by increasing the pressure again one can smoothly stay on branch β_1. The crucial observation is the drop from C to D is an irreversible path in the thermodynamical sense. That is, the drop from C to D is a sudden change in state* and should therefore not be considered a reasonable approximation to "a succession of equilibrium states traverse at an infinitesmally slow rate," i.e. a reversible path [13].

Let us now proceed to analyze the above two facts with thermodynamical language. To do so, introduce the free energy $F = E - TS$ where, by definition, E is the total internal energy of the gas and liquid in the cylinder and S is the entropy of the entire system in the cylinder.

By exterior differentiation, then,

$$dF = dE - TdS - SdT. \tag{4.5}$$

*A trivial remark should be made here is that any point on the branch β_2 and δ_2 represent a gas state and any point on the straight line $\overline{BA}$ represent a state with combination of liquid and gas.

We now consider an infinitesmal transformation of our system and apply to this transformation the first law of thermodynamics. We obtain

$$dE = dQ + dW \tag{4.6}$$

where Q is the heat flux and $dW = -PdV$ is the differential of the work performed by the external forces during a transformation. (In going from gas to liquid, one does $P - V$ work on the system, which, in going from liquid to gas, the system does work on its environment. Hence, the sign convention $dW = -PdV$ is correlated to this detail from reality.) Equation (4.6) can also be written as

$$E_B - E_A = (Q_B - Q_A) + (W_B - W_A) \tag{4.7}$$

along path γ in Figure 4.17 .

The second law of thermodynamics can be formulated as follows. If γ is any path, reversible or not, from state A to state B, then

$$\int_\gamma \frac{dQ}{T} \leq S_B - S_A , \tag{4.8}$$

with equality if and only if γ is reversible. An equivalent form could then be, if T is constant,

$$dQ \leq TdS . \tag{4.9}$$

If we put (4.9), (4.6) together in (4.5), we have

$$dF \leq -PdV . \tag{4.10}$$

In many thermodynamical transformations the pressure and the temperature of the system do not change. Under such circumstances it is possible to define a function G of the system which has the following property ("Gibbs free energy minimum principle"): if the function G is a minimum for a given set of values of the pressure and the temperature, then the system will be in equilibrium at the given P and T. The function G, is called Gibb's function or thermodynamical potential at constant pressure, is now defined as

$$G = F + PV .\qquad(4.11)$$

By exterior differentiation and with (4.10), we obtain

$$dG \leq 0 .\qquad(4.12)$$

Thus, if we consider an isothermal, isobaric transformation at the constant P and T which takes our system from a state A to a state B, we have

$$G(B) \leq G(A) .\qquad(4.13)$$

It follows from (4.13) that in an isobaric isothermal transformation of a system, the thermodynamic potential at constant pressure <u>can never increase</u>. We may therefore say that if T and P of a system are kept constant, <u>the state of the system for which G is a minimum is a state of stable equilibrium</u>. The reason for this is that if G is a minimum, any spontaneous change in the state of the system would have the effect of increasing G ; but this would be in contradiction to the inequality (4.13). Hence the above principle identifies the equilibrium states of a system (parametrized here by (V,P,T)) as nondegenerate minima of the corresponding map $G(-,P,T) = G_{(P,T)} \colon \mathbb{R} \to \mathbb{R}$. That is, at fixed pressure P and temperature T , the possible equilibrium positions of the system are nondegenerate minima of the Gibb's function at (P,T), considered only as a function of V . For the phase transition process therefore, the graph of the Van der Waals' isotherms give us the states and therefore the minima of the family of potential functions $\{G_{(P,T)}\}$. Thus, at P_τ, we have seen that $G_{(P_\tau,T)}(A) = G_{(P_\tau,T)}(B)$ where we now know that A and B (see Figure 4.18) are the only two nondegenerate minima of the map $G_{(P_\tau,T)} \colon \mathbb{R} \to \mathbb{R}$.

One more observation should be made here is that where should P_τ be located. This is known as Maxwell's (averaging) convention. Given a Van der Waals isothermal, we may now wish to determine that the pressure of the saturated vapor is when its temperature is equal to that of the given isothermal, or, geometrically speaking, how high above the V-axis we must draw the horizontal line BA which corresponds to the liquid-vapor state (see Figure 4.20). The averaging process takes place when the area BS_1K and KS_2A are the same. This fact can also be proved if we consider the reversibly isothermal cycle BS_1KS_2ADKB. The work performed during this cycle, as measured by its area, must vanish.

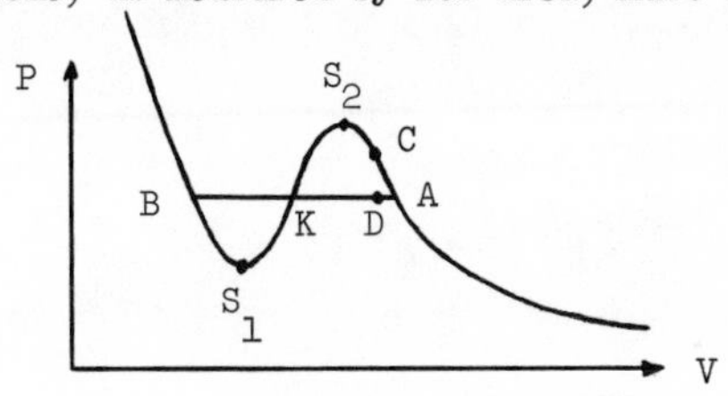

Figure 4.20.

Now we are ready to explain why the negative u-axis is the catastrophe set under Maxwell's convention. First of all, the Gibb's function is indeed the function in Figure 4.7 - 4.14, each one of these represents a state of the phase transition (the parameters u and v are now P, T respectively). Without loss of generality we choose c_0 as the gas state. By keeping T = constant, let P vary, each point on the branch β_2 is a stable equilibrium, in other words, is a state. (As a matter of fact, a gas state.) Until P reaches P_T, the corresponding figure of the potential G is Figure 4.11, we have two local minimum with the same value. From our mathematical model this is where the catastrophic situation happens and on the other hand from physics, this is exactly where the phase transition occurs. From our previous observation, as $c_0 = c_1$ or $G^{P_4}(V_1) = G^{P_4}(V_2)$, G reaches minimum, i.e. a state of stable equilibrium, then as indicated in Figure 4.18, the path AB will be followed. That is, the phase changing from gas to liquid occurs. From there on, as P varies further c_1 will be chosen as the stable equilibrium, in other words the phase transition has been done and it is liquid state from now on. This is why the choice of absolute minimum has been interpreted as Maxwell's convention by Thom.

Finally, in order to explain why Van der Waals' equation can be changed into the scheme of this example, we need the following clarification and coordinate transformation.

Clearly, as indicated already, the Gibb's function of the process "phase transition" given by $G(V,P,T)$ may be considered as an unfolding $G_{(P,T)}: \mathbb{R} \to \mathbb{R}$ of the map $G_{(P_0,T_0)}$ at an initial pair of parameter values for which the map $G_{(P_0,T_0)}$ has a degenerate singularity. From Van der Waals' equation for one mole of gas

$$(P + \frac{a}{V^2})(V - b) = rT , \tag{4.14}$$

one finds that $(P_0,T_0) = (\frac{a}{27b^2}, \frac{8a}{27br})$.

Moreover, we can use Thom's classification theorem to write G (as a function of one variable with two parameters) in a particularly simple form. Of the seven Thom polynomials (in the list), only one has unfolding codimension equal to two, $P(x) = \frac{1}{4}x^4$. Thus, as a very reasonable assumption, we can hypothesize that G is right equivalent to a reparametrization $h: \mathbb{R}^2 \to \mathbb{R}^2$, sending $(P,T) \to (u,v)$, of the universal unfolding of $P(x)$, namely $f(x,u,v) = \frac{1}{4}x^4 + \frac{1}{2}ux^2 + vx$.

Since we have an explicit description of the state space via Van der Waals' equation, we can construct h as follows.

If $(\frac{a}{27b^2}, \frac{8a}{27br})$ are to be the initial parameter coordinates of the singularity map, we can transform them to $(1,1)$ by setting

$$\begin{aligned} \overline{P} &= \frac{27b^2}{a} P \\ \overline{T} &= \frac{27br}{8a} T . \end{aligned} \tag{4.15}$$

We can also normalize V. At $(\frac{a}{27b^2}, \frac{8a}{27br})$, the critical point has V value $3b$. Thus, set $\overline{V} = \frac{V}{3b}$.

Now

$$(P + \frac{a}{V^2})(V - b) = (\frac{a}{27b^2}\overline{P} + \frac{a}{9b^2\overline{V}^2})(3b\overline{V} - b)$$

$$= 3b\frac{a}{27b^2}(\overline{P} + \frac{3}{\overline{V}^2})(\overline{V} - \frac{1}{3})$$

$$= \frac{a}{9b}(\overline{P} + \frac{3}{\overline{V}^2})(\overline{V} - \frac{1}{3})$$

$$= \frac{8a}{27br} r\overline{T} .$$

That is,

$$(\overline{P} + \frac{3}{\overline{V}^2})(\overline{V} - \frac{1}{3}) = \frac{8}{3}\overline{T} \tag{4.16}$$

is another form of Van der Waals' equation.

With one mole of gas, $\overline{X} = \frac{1}{\overline{V}}$ measures the (absolute) density of the gas-liquid (up to the constant 3b) in the cylinder in its own phase.

Introducing this variable into (4.16), we get

$$(\overline{P} + 3\overline{X}^2)(\frac{1}{\overline{X}} - \frac{1}{3}) = \frac{8}{3}\overline{T}. \tag{4.17}$$

(1,1,1) in $(\overline{X},\overline{P},\overline{T})$ coordinates is still the critical point. We can shift it to (0,0,0) by introducing

$$\begin{aligned} p &= \overline{P} - 1 \\ x &= \overline{X} - 1 \\ t &= \overline{T} - 1 \ . \end{aligned} \tag{4.18}$$

In (x,p,t) coordinates, (4.16) takes the form

$$x^3 + \frac{1}{3}(8t + p)x + \frac{2}{3}(8t - 2p) = 0 \tag{4.19}$$

Thus, letting

$$\begin{aligned} u &= \frac{1}{3}(8t + p) \\ v &= -\frac{2}{3}(4t - p) \end{aligned} \tag{4.20}$$

which defines our reparametrization map h .

One conclusion that can be drawn is the relationship between the phase transition pressure P_T and the temperature. For in the (u,v) coordinates, the condition that $f_{(u,v)}$ have two stable minima for which $f_{(u,v)}(x)$ has the same critical value is $v = 0$, $u < 0$ (noticing that this is exactly the negative u-axis). That is

$$\begin{aligned} 4t - p &= 0 \\ 8t + p &< 0 . \end{aligned} \tag{4.21}$$

So, working backwards, we get

$$\begin{aligned} 0 = 4t - p &= 4\overline{T} - \overline{P} - 3 \\ &= 4\left(\frac{27br}{8a}\right)T - \left(\frac{27b^2}{a}\right)P - 3 \\ &= \frac{27br}{2a}T - \frac{27b^2}{a}P - 3 . \end{aligned}$$

Hence,

$$P_T(T) = \frac{1}{27b^2}\left(\frac{27br}{2a}T - 3\right)a = \frac{r}{2b}T - \frac{a}{9b^2} . \tag{4.22}$$

This is by the way, a testable conclusion. One should be aware of the experimentally determined observation that Van der Waals' equation adequately describes the (P,V,T) inter-relationship sufficiently far away from the critical point. On the other hand, Thom's classification theorem would apply sufficiently near the critical point to the representatives of the germs of G and f at the critical point. Thus, there is a formalistic subtlety here which seemingly should generate caution in going directly from the simplistic mathematics to the complex reality. Nonetheless, as a model of this (and other types as well) phase transition, which incorporates Berthelot's equation [84] and presumably any other empirically derived mathematical model of phase change, Thom's systematization offers, if not a scientifically theoretical explanation ("predicative explanation") at least a mathematically elegant and (topologically) universal one.

CHAPTER 5

THOM-WHITNEY STRATIFICATION THEORY

1. Introduction

Stratification is one of the fundamental concepts in algebraic geometry and the theory of singularities. The study of stratifications originated with the work of Whitney and Thom ([71], [89]) on singularities of analytic varieties. Our purpose, in this chapter is to give a brief exposition of Thom-Whitney stratification theory. It is not our intention to provide the details in the development ([36,49,50,65,78, 90]) of this theory. We do not prove those fundamental theorems stated in section 4. Our primary concerns in this chapter are a detail study of Whitney's regularity conditions and how to decide whether a stratification of a variety is regular. Many examples have been provided to illustrate these points.

The objects to be studied in this chapter will be varieties - algebraic, analytic or semi-analytic.

Definition 1.1. A real (or complex) algebraic variety V is a point set in $\mathbb{R}^n$ ($\mathbb{C}^n$, respectively) which is the set of common zeros of a set of polynomials.

In the study of functions of several variables, as in the study of functions of a single variable, the set of zeros of a function (or, in higher dimensional cases, the set of common zeros of a set of functions) plays a very important role. In the one-dimensional case, the geometrical situation is simple, since such zero sets must be discrete. In the case of several variables, the situation is far more complicated. Submanifolds are defined (locally) as zero sets of certain functions, but not all zero sets are submanifolds (see the following examples). Our primary interest in this chapter is with those varieties which are not submanifolds.

Example 1.1. $f(x,y) = x^2 - y^2$. The set of zeros of f,

$V = \{(x,y) \in \mathbb{R}^2 \mid x^2 = y^2\}$, is a variety. It is not a manifold, since the origin separates it into four components (into two components in case $(x,y) \in \mathbb{C}^2$), see Figure 1.1, and hence a neighborhood of the origin is not homeomorphic to $\mathbb{R}^2$ ($\mathbb{C}^2$ respectively).

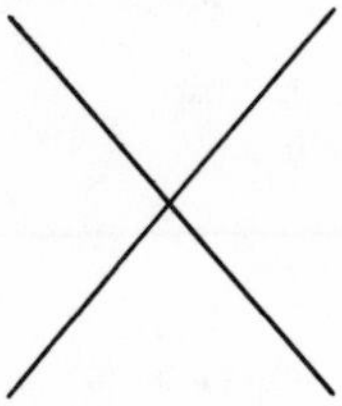

Figure 1.1.

Example 1.2. $f(x,y,z) = z^2 - xy$. The set $V = \{(x,y,z) \mid z^2 = xy\}$ is the variety of f, see Figure 1.2.

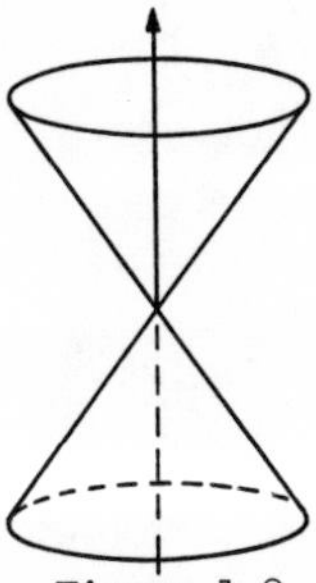

Figure 1.2.

In this example, the Jacobian matrix of f is $(-y,-x,2z)$. Thus at every point except the origin, V is locally a submanifold, but V itself is not a manifold.

Similarly to Definition 1.1, we have:

Definition 1.2. A real (or complex) analytic variety is the set of common zeros defined by a finite number of analytic functions.

In connection with analytic inequalities we make the following definitions:

Definition 1.3. A basic semi-analytic set-germ at x_0 is the germ of the set at x_0 defined by a finite number of analytic inequalities, namely $|f_i(x)| \geq 0$, $1 \leq i \leq s$, where the f_i's are analytic.

A set S is semi-analytic if at every $x_0 \in S$, the germ of S at x_0 is the finite union of set-germ of the form U - V , where U, V are basic semi-analytic set-germs [35].

The basic idea in stratification is to decompose a variety V (algebraic, analytic or semi-analytic) into a finite, disjoint union of open or closed manifolds without boundary (just as with simplicial or CW complexes),

$$V = M_1 \cup M_2 \cup \dots \cup M_s ,$$

in such a way that each manifold M_i consists of "equally bad" points. Each M_i is called a stratum of the stratification. The concept of "equal badness" will be defined in section 3 through regularity conditions and it will be explained naively by the examples in the next section.

2. Examples

Example 2.1. $f_1(x,y,t) = t(x^2 - y^2) + x^4 + y^4$,

$$V_1 = \{(x,y,t) \mid t(x^2 - y^2) + x^4 + y^4 = 0\} . \tag{2.1}$$

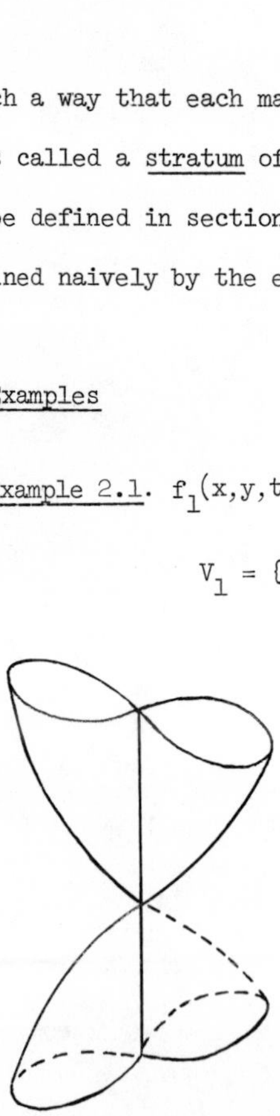

Figure 2.1.1.

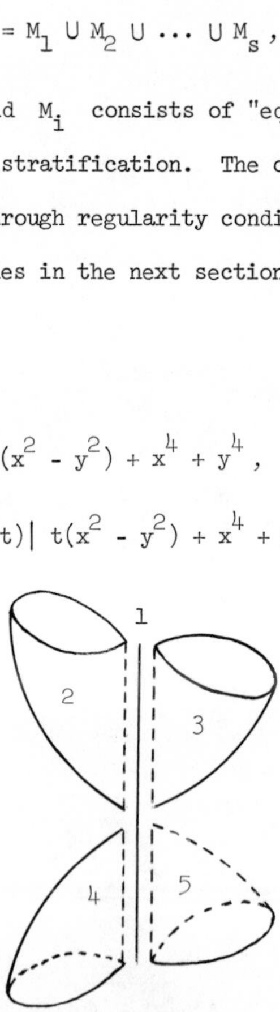

Figure 2.1.2.

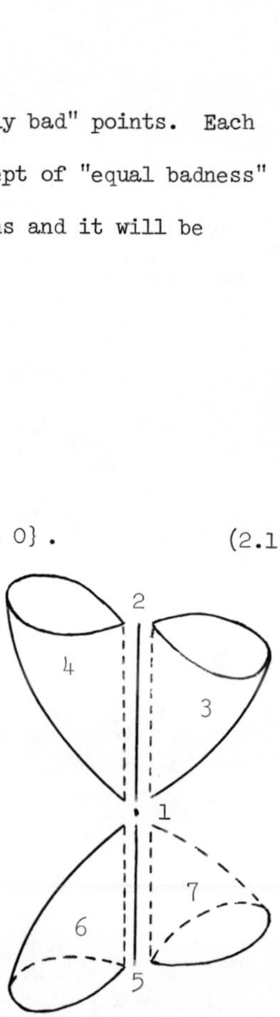

Figure 2.1.3.

Example 2.1. $f_2(x,y) = x(x - y^2)$

$$V_2 = \{(x,y) \mid x(x - y^2) = 0\} \tag{2.2}$$

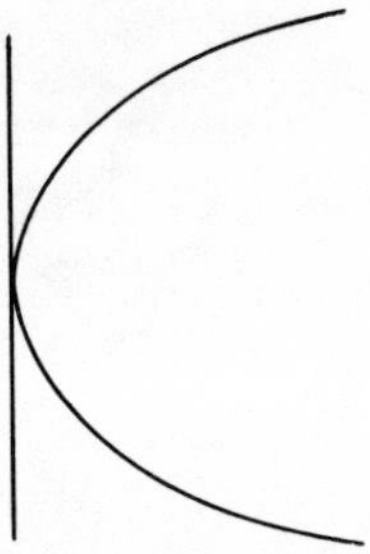

Figure 2.2.1.

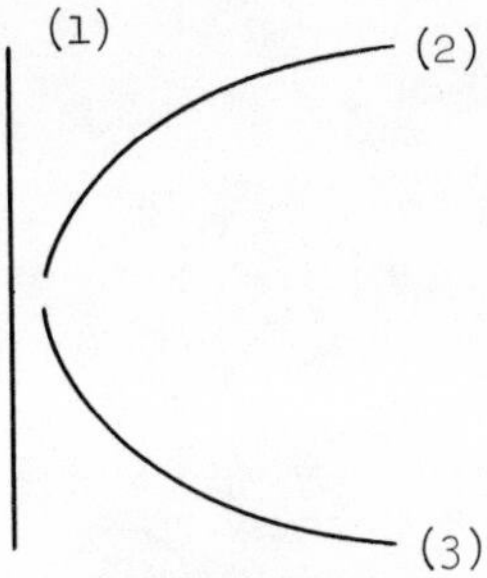

Figure 2.2.2.

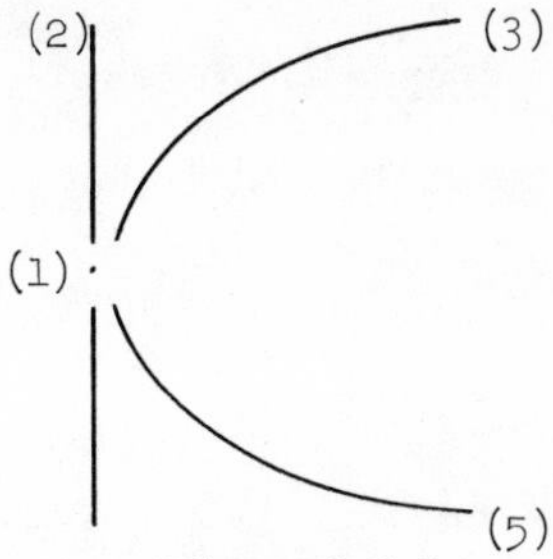

Figure 2.2.3.

<u>Example 2.3</u>. $f_3(x,y,t) = y^2 - t^2x^2 - x^3$

$$V_3 = \{(x,y,t) \mid y^2 - t^2x^2 - x^3 = 0\} \tag{2.3}$$

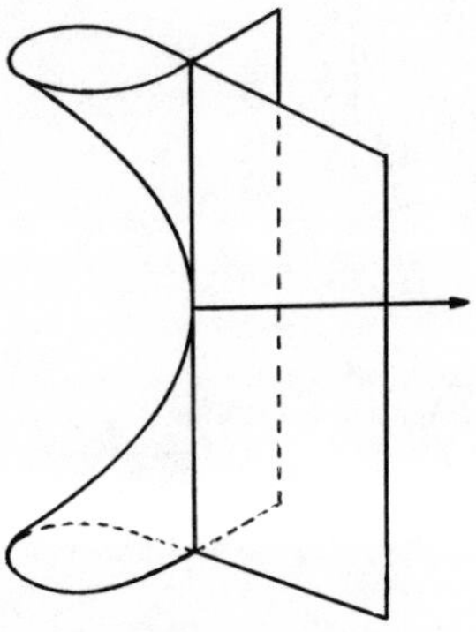

Figure 2.3.1.

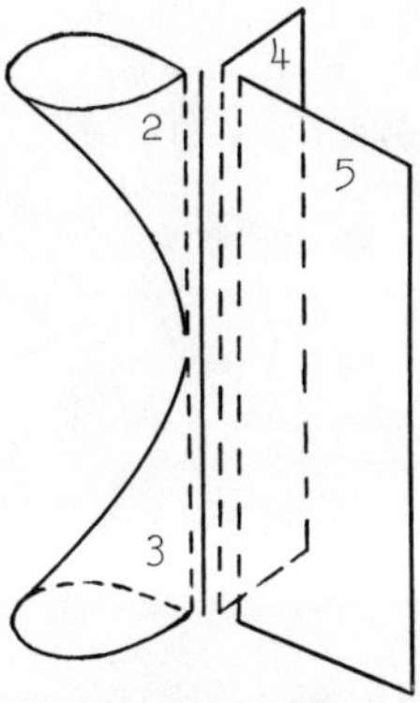

Figure 2.3.2.

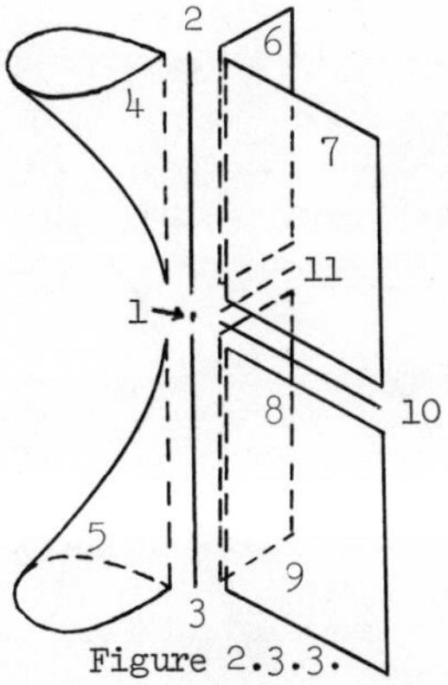

Figure 2.3.3.

<u>Example 2.4</u>. $f_4(x,y,t) = y^2 - t^2x^3 - x^5$

$$V_4 = \{(x,y,t) \mid y^2 - t^2x^3 - x^5 = 0\} \tag{2.4}$$

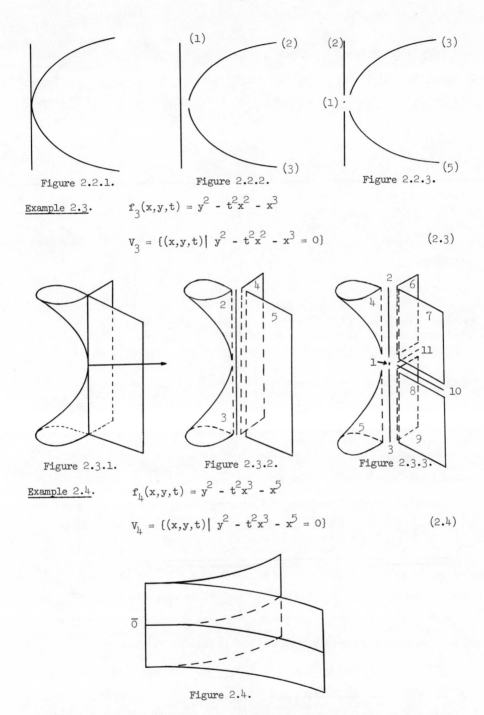

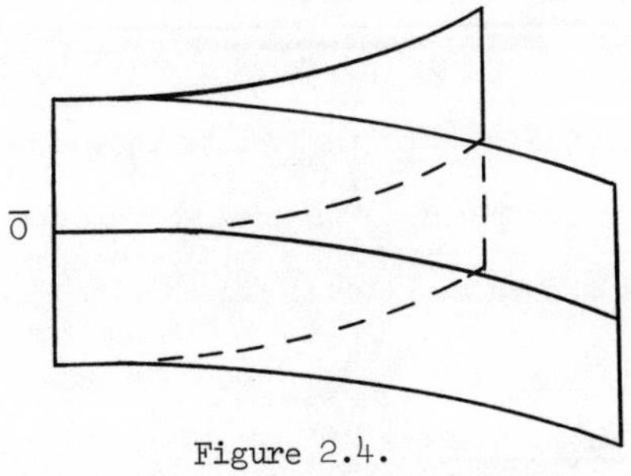

Figure 2.4.

The variety in this example looks like two sheets of paper which have been glued and pinched together along one edge and then pinched harder at $\overline{0} \in \mathbb{R}^3$.

Example 2.5. The slow spiral.

$$V_5 = \left\{ (r,\theta) \,\middle|\, r = \begin{cases} e^{-\theta} & \\ 0 & \text{as } \theta \to \infty \end{cases} \right\}$$

Figure 2.5.1.

Figure 2.5.2.

Example 2.6. The quick spiral.

$$V_6 = \left\{ (r,\theta) \,\middle|\, r = \begin{cases} e^{-\theta^2} & \\ 0 & \text{as } \theta \to \infty \end{cases} \right\}$$

Figure 2.6.1.

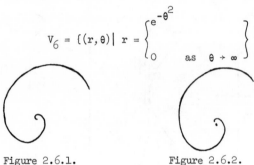

Figure 2.6.2.

Examples 2.1, 2.2 and 2.3 will be discussed here, although Example 2.3 is actually just a combination of Example 2.1 and Example 2.2 . For each of these three examples, we have two kinds of stratification as indicated by Figure 2.1.2 and Figure 2.1.3, Figure .2.2 and Figure 2.2.3, etc. Intuitively Figures 2.1.2, 2.2.2 and 2.3.2 are not what we want in the sense that, as we have said, each stratum should consist of "equally bad" points. However the origin in stratum 1 in each case is "worse" than the rest of the points in the same stratum. To make the words "bad" and "worse" more precise, we need to describe some conditions which govern how the strata should be patched together.

3. Regularity Conditions of H. Whitney

Let $V = M \cup P \cup \ldots$ be a variety in $\mathbb{R}^n$ where M, P, ... are manifolds without boundary. Let $p \in P$, and let $T(P,p)$ be the tangent space of P at p . Let $m \in M$, and let $N(M,m)$ be the normal space of M at m . We will denote by π some appropriate projection: for instance, $\pi_{N(M,m)}(\tau)$ is the projection of τ into the normal space of M at m.

<u>Definition 3.1</u>. M is <u>(a)-regular</u> over P at p if, for any tangent vector $\tau \in T(P,p)$,

$$\lim_{m \to p} \pi_{N(M,m)}(\tau) = 0 \quad . \tag{3.1}$$

This means that for $\tau \in T(P,p)$, the projection of τ into the normal space of M at m approaches zero as m approaches p. That is, τ is nearly perpendicular to N(M,m) when m is close to p; equivalently, τ is nearly contained in T(M,m) when m is close to p. Thus we may restate the condition that M be (a)-regular over P at p in the form: <u>T(P,p) is nearly contained in T(M,m) as m approaches p</u>. We could interpret $|\pi_{N(M,m)}(\tau)|$ as the distance from τ to T(M,m). Now let us check the first two examples in section 2.

<u>Example 2.1</u>. $V_1 = \{(x,y,t) \mid t(x^2 - y^2) + x^4 + y^4 = 0\}$.

(1) Let us look at the stratification shown in Figure 2.1.2 first.

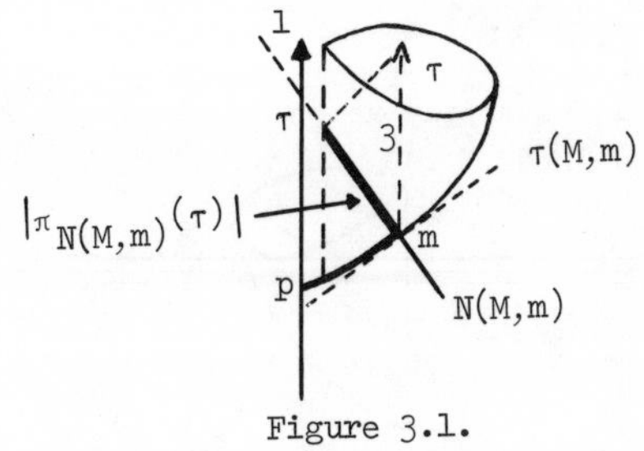

Figure 3.1.

Let m approach p as shown in Figure 3.1; it is clear that T(M,m) is far from containing τ. It is also quite obvious that $|\pi_{N(M,m)}(\tau)|$ does not approach zero as m approaches p. This indicates that the stratification of V_1 via Figure 2.1.2 does not satisfy the (a)-regularity condition, or, as we may say, is not stratified according to the (a)-regularity condition. In particular (3) is not (a)-regular over (1) at p.

(2) Next, let us look at the stratification of V_1 via Figure 2.1.3, namely

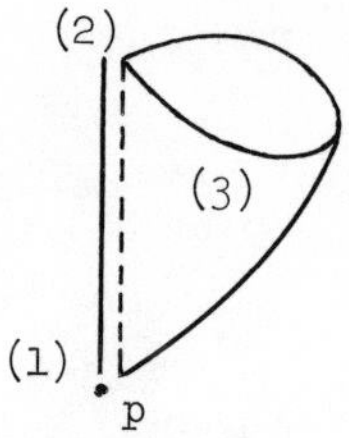

Figure 3.2.

To check whether the stratification as shown in Figure 3.2 is (a)-regular, there are several cases to be looked at in detail.

(i) Is (2) (a)-regular over (1) at p ?

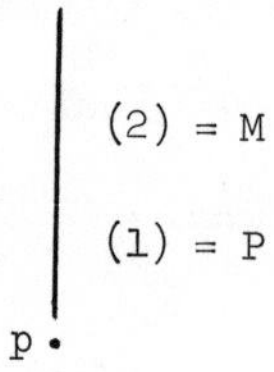

Figure 3.3.

This is a trivial case since (1) is the point p, and in particular $T(P,p)$ (i.e. the point p in this case) is clearly contained in $T(M,m)$ for every $m \in M$.

(ii) Is (3) (a)-regular over (1) at p ?

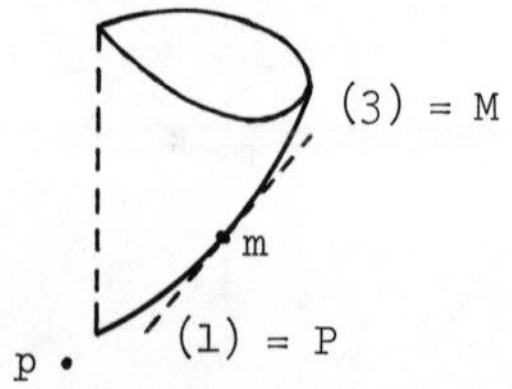

Figure 3.4.

This is also an obvious case since $T(P,p)$ (i.e. the point p in this case) is nearly contained in $T(M,m)$ as m approaches p.

(iii) Is (3) (a)-regular over (2) at any point p in (2) ?

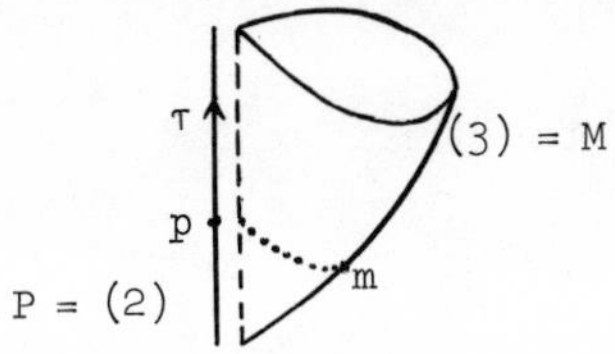

Figure 3.5.

Let us take any sequence of points on M such that $m \to p$ (as indicated in the dotted line of Figure 3.5). The tangent space $T(M,m)$ is a two-dimensional linear space; thus it is clear that τ is nearly contained in $T(M,m)$ as $m \to p$. Hence (3) is (a)-regular over (2) at any p in (2), i.e. (3) is (a)-regular over (2).

Moreover (i), (ii) and (iii) exhaust all possibilities in our checking process, therefore the stratification in Figure 3.2 (or Figure 2.1.3) satisfies the (a)-regularity condition.

<u>Example 2.2.</u> $V_2 = \{(x,y) \mid x(x - y^2) = 0\}$.

(1) In this example, let us consider the stratification as shown in Figure 2.2.3 first, namely

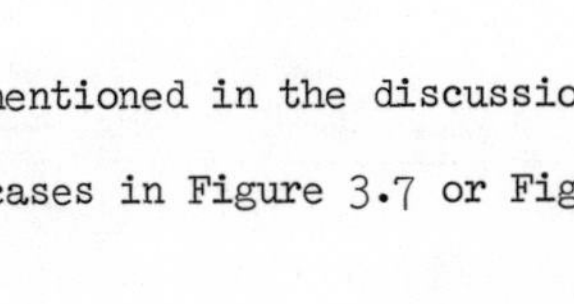

Figure 3.6.

For the same reason as we mentioned in the discussion of (i) in Example 2.1, it is clear that either of the cases in Figure 3.7 or Figure 3.8 is (a)-regularly

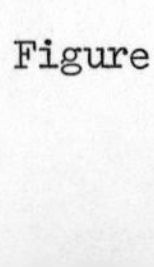

Figure 3.7.

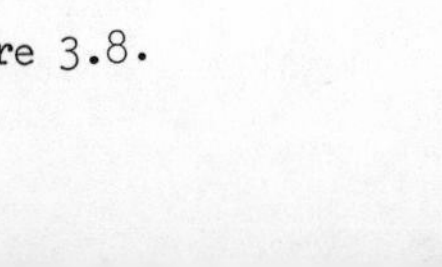

Figure 3.8.

stratified. Thus Figure 2.2.3 is an (a)-regular stratification.

(2) Next let us consider the stratification shown in Figure 2.2.2 .

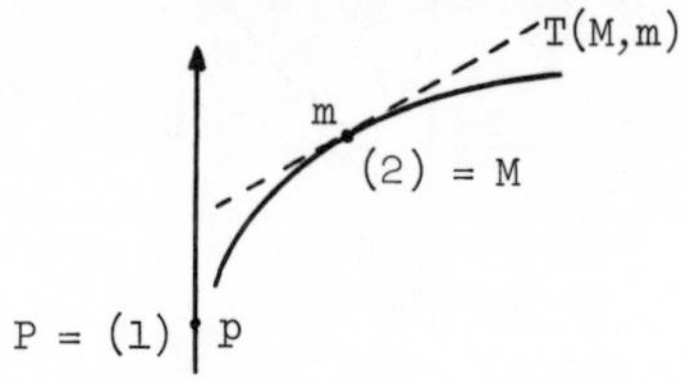

Figure 3.9.

The question is whether (2) is (a)-regular over (1) at the origin p . Unfortunately the stratification in this case does satisfy the (a)-regularity condition, since τ is clearly contained in $T(M,m)$ as $m \to p$. Here we say "unfortunately," because p in the stratum P is obvious "worse" than the other points in P .

Therefore we need further regularity conditions. Before we state the regularity conditions (b') and (b) let us fix some notation.

Let $p \in P$, $m \in M$. Denote the vector $\overrightarrow{mp}$ by $\overline{m}$, $\pi_{T(P,p)}(\overline{mp})$ by $\pi_p(\overline{m})$. Let $\{p_i\}$ be a sequence of points in P, $\{m_i\}$ a sequence of points in M. Denote the vectors $\overrightarrow{m_i p_i}$ by $\overrightarrow{m_i}$, $\pi_{T(P,p_i)}(\overline{m_i p_i})$ by $\pi_{p_i}(\overline{m}_i)$.

<u>Definition 3.2</u>. M is (b')-regular over P at p if

$$\lim_{m \to p} \pi_{N(M,m)} \frac{\overline{m} - \pi_p(\overline{m})}{|\overline{m} - \pi_p(\overline{m})|} \tag{3.2}$$

where $|\cdot|$ means the Euclidean norm.

Intuitively, this means that $T(M,m)$ nearly contains the <u>direction</u> of the vector $\overline{m} - \pi_p(\overline{m})$; in other words, $N(M,m)$ is nearly perpendicular to the direction $\overline{m} - \pi_p(\overline{m})$.

<u>Definition 3.3</u>. M is (b)-regular over P at p if

$$\lim_{i \to \infty} \pi_{N(M,m_i)} \frac{\overline{m}_i - \pi_{p_i}(\overline{m}_i)}{|\overline{m}_i - \pi_{p_i}(\overline{m}_i)|} = 0 . \tag{3.3}$$

Remark. J. Mather proved that the regularity conditions (a) and (b) are equivalent to the regularity conditions (a) and (b'). He also proved that (b) implies (a); thus it is enough to check the condition (b). However condition (b') is obviously easier to handle than (b) and the condition (a) is not difficult to check at all. Therefore we prefer to consider the conditions (a) and (b'). The proof of the equivalence of (a) + (b') with (a) + (b) will be given after we have demonstrated condition (b') with the above mentioned examples.

Definition 3.4. A stratum M of the variety V is regular over a stratum of V if M is (a)-,(b)-regular (or (a)-,(b')-regular) over P at every $p \in P$. (These two conditions have often been referred to as Whitney's regularity conditions.)

Definition 3.5. A stratification of a variety is regular if for any two strata M, P of V with $P \subset \overline{M}$, M is regular over P.

Recall that our intention in giving Definition 3.2 (or Definition 3.3) was to distinguish the origin in the stratum (1) of V_2 in Figure 2.2.2. Therefore let us look at this example first.

(i) Consider Figure 2.2.2, or Figure 3.10:

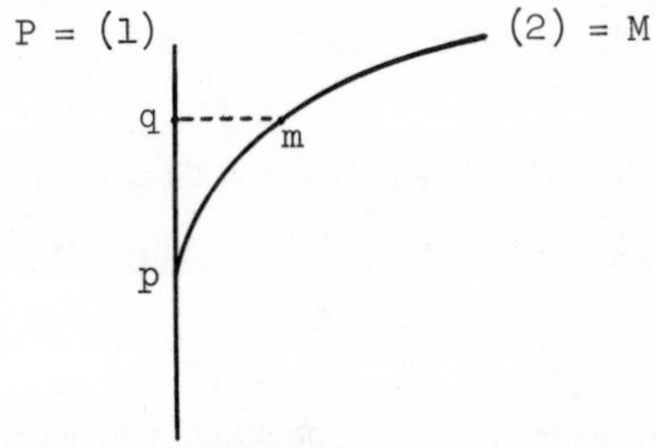

Figure 3.10.

Let $m \in M$ and project $\overline{mp}$ onto P since $P = T(P,p)$ in this case. Then $\overline{mq} = \overline{m} - \pi_p(\overline{m})$. It is clear that the direction (or unit vector) of $\overline{mq}$ is not nearly contained in $T(M,m)$ as $m \to p$, which violates (b'). Thus the stratification of $V_2 = \{(x,y) \mid x(x - y^2) = 0\}$ in Figure 2.2.2 is not regular.

(ii) Consider Figure 2.2.3, or Figure 3.11:

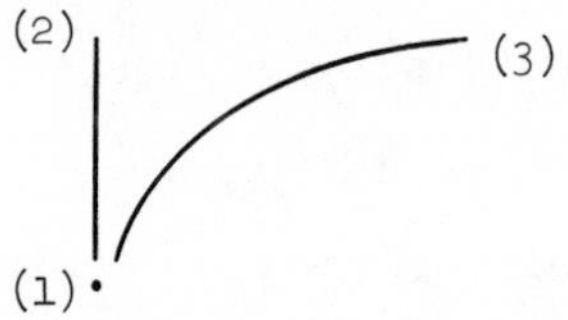

Figure 3.11.

It is clear that (2) is regular over (1) and (3) is regular over (1) since (1) is just a point, and further $(2) \not\subset \overline{(3)}$ nor $(3) \subset \overline{(2)}$, thus V_2 is regularly stratified via Figure 2.2.3.

For the same reason as in (i), it is obvious that V_1 is not (b')-regularly stratified via Figure 2.1.2. (In general, if a stratification is not (a)-regular, we do not proceed to check the regularity condition (b').) However V_1 is regularly stratified via Figure 2.1.3, or the following Figure 3.12. As usual, it is trivial to see that both (2) and (3) are regular over (1) since (1)

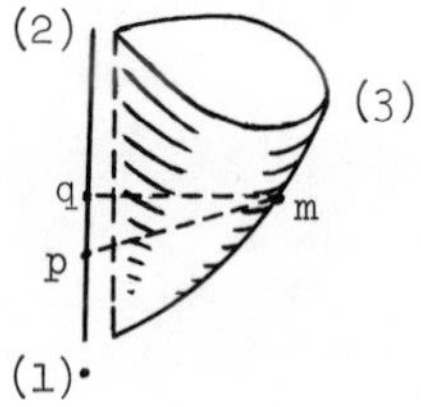

Figure 3.12.

is a point. We only have to check that (3) is (b')-regular over (2). Let p be any point in $(2) = P$ and $m \in (3) = M$. Take any sequence of points in M such that $m \to p$. It is not difficult to visualize that as m approaches p in (2) the line in the direction $\overline{mq} = \overline{m} - \pi_p(\overline{m})$ is nearly contained in the (2-dimensional) tangent space $T(M,m)$ because the direction $\overline{mq}$ is revolving about (2). This proves that (3) is (b')-regular over (2) and hence V_1 is regular via the stratification shown in Figure 2.1.3. V_3, defined in section 2 could serve as a good exercise for the reader to check the regularity conditions.

Remark. After the discussion of these examples, it is reasonable to expect (actually proved by H. Whitney) that if V is stratified into $M_1 \cup M_2 \cup \cdots \cup M_s$ and if M_j is (a)-,(b)-regular over M_i such that $M_i \cap \overline{M}_j = \emptyset$, then

$$\dim M_i < \dim M_j . \qquad (3.4)$$

In general (3.4) is not necessarily true. For example, let $M_1 = \{0\} \times [-1,1]$ in $\mathbb{R}^2$ and $M_2 = \{(x,y) \mid y = \sin \frac{1}{x},\ x \neq 0\}$. We have $M_1 \cap \overline{M}_2 \neq \emptyset$ but $\dim M_1 = \dim M_2$ (see Figure 3.13).

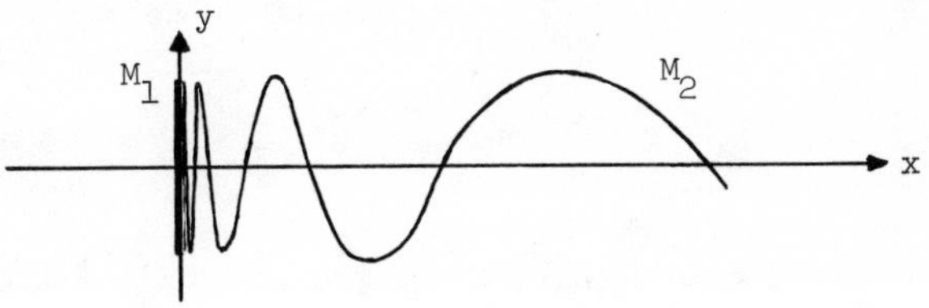

Figure 3.13.

R. Thom conjectured in 1970 that if $V = M_1 \cup \cdots \cup M_s$ is a (a)-,(b)-regular stratification and $M_i \cap \overline{M}_j \neq \emptyset$ then $M_i \subset \overline{M}_j$. Mather proved this statement in his notes on Topological Stability [47].

This leads us to the following definition:

Definition 3.6. A stratification $V = M_1 \cup \cdots \cup M_s$ satisfies the frontier condition if whenever $M_i \cap \overline{M}_j \neq \emptyset$ then $M_i \subset \overline{M}_j$.

Notice that the stratification of V_2 via Figure 2.1.2 does satisfy the regularity condition (a) but fails to satisfy the frontier condition (since $P \cap \overline{M} \neq \emptyset$ but $P \not\subset \overline{M}$), thus by Thom-Mather Theorem (see next section), the statification of V_2 via Figure 2.1.2 is not regular.

Finally, we will conclude this section by proving the equivalence of (a) + (b) with (a) + (b'). It suffices to prove that (a) + (b') implies (a) + (b) since the converse implication is clear.

Theorem 3.1. Let M, P be two strata of a stratification of a variety (algebraic, analytic or semi-analytic) such that $P \subset \overline{M}$.

Proof: Let p be any point of P. We claim that M is (b)-regular over P at p under the given hypothesis. Since $P \subset \overline{M}$ and $P \cap M = \emptyset$, $P \subset \partial M$, the boundary of M. Let $\{p_i\}$ be a sequence of points in P and $\{m_i\}$ a sequence of points in M such that $p_i \to p$ and $m_i \to p$ as $i \to \infty$. Locally near p, P looks like $\mathbb{R}^k \times \{\overline{0}\}$ and M looks like $\mathbb{R}^n$. Let π be the projection of M onto $T(P,p)$ in a neighborhood of p. Denote $\pi(p_i)$ by q_i, then $q_i \to p$.

Consider the line segments $\ell_i^{(1)} = \overrightarrow{m_i q_i}$, $\ell_i^{(2)} = \overrightarrow{q_i p_i}$ and $\ell_i^{(3)} = \overrightarrow{m_i p_i}$, they all lie in an n-dimensional Euclidean space $\mathbb{R}^n$. We can write

$$\ell_i^{(3)} = \ell_i^{(1)} + \ell_i^{(2)} . \tag{3.5}$$

This is an equation about one-dimensional subspaces in $\mathbb{R}^n$. If the affine spaces $\ell_i^{(1)}$, $i = 1, 2, 3$, are transferred to linear spaces by sending q_i to the origin, then (3.5) is a precise relation between linear spaces.

Now, since $\ell_i^{(1)} \to \ell^{(1)}$ and $\ell_i^{(2)} \to \ell^{(2)}$, it follows that $\ell_i^{(3)} \to \ell^{(1)} + \ell^{(2)}$. By the hypothesis of the theorem, if $T(M,m_i) \to \tau$ as $i \to \infty$, the regularity condition (b') shows that $\ell^{(1)} \subset \tau$. Moreover, $\ell^{(2)} \subset T(P,\overline{0}) = \mathbb{R}^k \times \{0\}$, so $\ell^{(2)} \subset T(P,\overline{0}) \subset \tau$ by regularity condition (a). Hence $\ell^{(1)} + \ell^{(2)} = \ell^{(3)} \subset \tau$ and condition (b) holds.

4. Fundamental Theorems

In this section we make an attempt to state the theorems which are most relevant in answering natural questions of the following kind:

(1) Given a variety, does it admit a Whitney regular stratification?

(2) Do the stated Whitney regularity conditions justify the claim made in section 1, namely, that each stratum consists of "equally bad" points?

(3) How does the Whitney stratification help us to study the local structure of a smooth mapping? (This is, of course, one of the fundamental objectives of the theory of the singularities of smooth mappings.)

(I) Existence Theorems.

Theorem 4.1. (Whitney, 1965 [90]) Every local analytic variety admits an (a)-,(b)-regular stratification.

Theorem 4.2. (Łojasiewicz, 1965 [35]) Every local semi-analytic variety admits an (a)-,(b)-regular stratification.

(II) Justification of the Regularity Conditions.

Theorem 4.3. (Thom-Mather [49, 71]) Let V be statified under the (a)-,(b)-regularity conditions. Along each stratum M_i the local topological picture remain invariant in the following sense: if x, y are two points in a connected component of the stratum M_i, then there is a neighborhood U_x of x in V, homeomorphic to a sufficient small neighborhood U_y of y in V, and the homeomorphism h preserves the stratification, i.e.

$$h: M_j \cap U_x \to M_j \cap U_y$$

is a homeomorphism for each j.

Example 4.1. Same as Example 2.1 (see Figure 4.1 below). Locally at x and

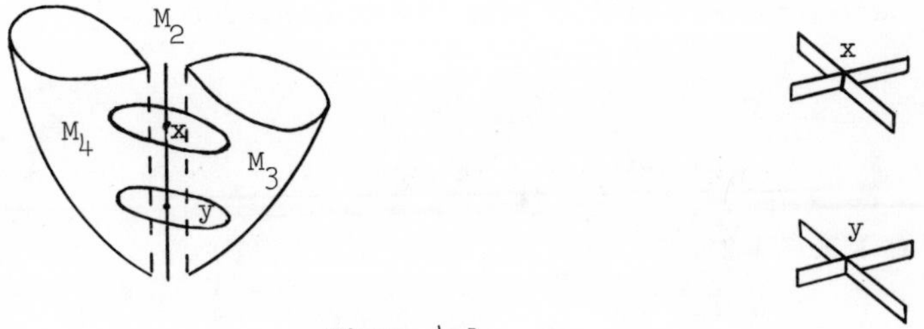

Figure 4.1.

y, $U_x \cap M_i$ is homeomorphic to $U_y \cap M_i$ for $i = 2, 3, 4$.

This theorem is important because as we have said, the fundamental idea in the notion of stratification is to decompose a variety into a disjoint union of strata each of which consists of equally bad points. This theorem justifies the claim that the Whitney regularity conditions satisfy this intuitive demand.

III. Isotopy Theorems.

The most general problem in Local Differential Analysis may be stated as follows: Given a C^m-mapping $G: U \to \mathbb{R}^p$, $G = (G_1,\ldots,G_p)$, where U is an open subset in $\mathbb{R}^n$, let $A = \{x \in U \mid G_i(x) = 0,\ i = 1,\ldots,p\}$. What can be said of the topological structure of A? In case G does not have singularities, the important Morse Theorem [51] answers our question in the following way:

Theorem 4.4. (Morse) Let M be a compact differentiable manifold and $f: M \to \mathbb{R}$ a differentiable function on M. If $df(p) \neq 0$ for all points $p \in M$ with $a \leq f(p) \leq b$, then $f^{-1}(a)$ is homeomorphism to $f^{-1}(b)$.

Isotopy Theorems are the generalizations of Theorem 4.4 in dealing with varieties. In the following, the maps under consideration will all be differentiable in an open set of an ambient (Euclidean) space containing a variety V.

Theorem 4.5. (First Isotopy Lemma [47] [71]) Let V be a (a)-,(b)-regularly stratified variety. Let $f: V \to \mathbb{R}$ be a proper map (i.e. inverse images of compact sets are compact) such that, for each stratum M_i of V, $f|_{M_i}$ maps M_i onto the open interval $(0,1)$ and $f|_{M_i}$ is a submersion. Then $f^{-1}(\alpha)$ is homeomorphic to $f^{-1}(\beta)$ for $\alpha,\beta \in (0,1)$.

Remarks.

(1) The requirement that $f|_{M_i}$ be a submersion avoids the situation indicated in Figure 4.1.

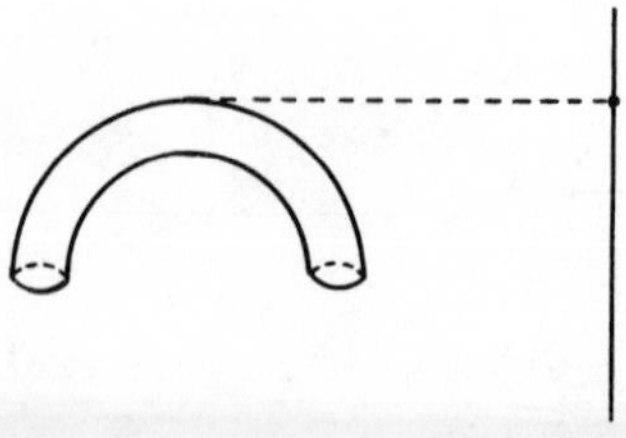

Figure 4.2.

(2) This theorem is important in the sense that it gives us conditions on the variety and on the real-valued functions such that the locus of f at the

values α and β are topologically equivalent (i.e. homeomorphic). More importantly, we may ask under what conditions will the <u>two mappings have the same topological type</u>. This question will be answered by the Second Isotopy Lemma, which gives an analogous result for mappings instead of loci.

To discuss the Second Isotopy Lemma, we first consider the following definitions, which are essentially due to Mather [47].

<u>Definition 4.1.</u> Consider a diagram of spaces and mappings:

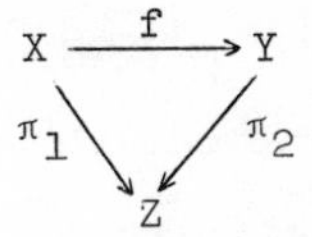

Diagram 4.1.

The mapping f is <u>trivial</u> over Z if there exist spaces X_0 and Y_0, a mapping $f_0: X_0 \to Y_0$ and homeomorphisms $X \tilde{\to} X_0 \times Z$, $Y \simeq Y_0 \times Z$ such that the following diagram of spaces and mapping is commutative:

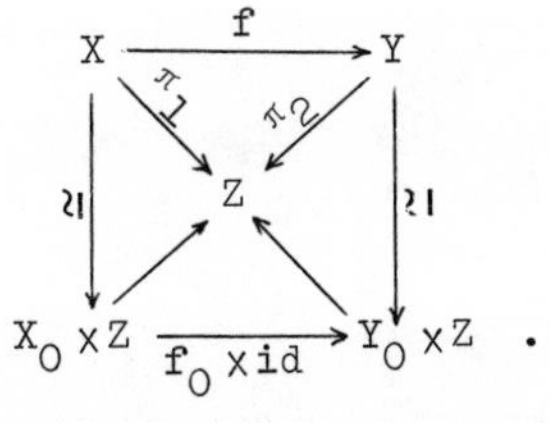

Diagram 4.2.

We say f is <u>locally trivial</u> over Z if for any $z \in Z$, there is a neighborhood U of z in Z such that in the diagram f is trivial over U.

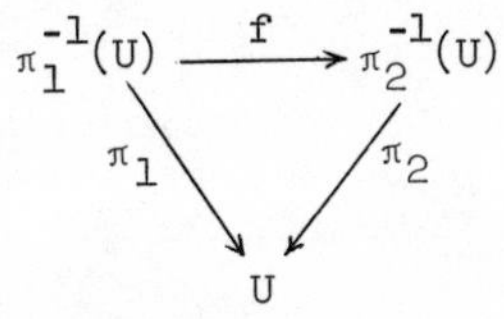

Diagram 4.3.

Local triviality of a mapping f over a space Z (we will take Z to be the open interval $(0,1)$ in our case) has, a consequence which will be very important in what follows. We think of f as a family $\{f_\alpha : \alpha \in (0,1)\}$ of mappings, where $f_\alpha : X_\alpha \to Y_\alpha$ is the mapping obtained by restricting f to the fibre X_α of X over α. If f is locally trivial over Z, then for any α, β in $(0,1)$, the mappings f_α and f_β are equivalent in the sense that there exist homeomorphisms $h: X_\alpha \to X_\beta$ and $h': Y_\alpha \to Y_\beta$ such that $h'f_\alpha = f_\beta h$ (in fact, this is the same equivalence relation as is used in the definition of topologically stable mappings in section 4 of Chapter 1).

Now suppose that A, A' are smooth manifolds and that V, V' are varieties in A, A' respectively; and suppose further that V and V' are both Whitney regularly stratified. Let $g: A' \to A$ be a smooth mapping and suppose $g(V') \subseteq V$. Thom's second isotopy lemma gives sufficient conditions for the following diagram to be locally trivial:

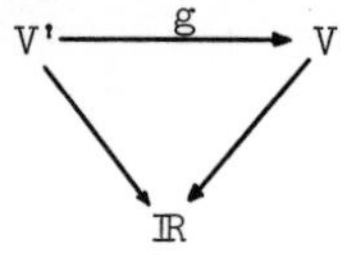

Diagram 4.4.

To state Thom's second isotopy lemma, we must introduce Thom's condition a_g. Let M and P be submanifolds of A' and let p be a point in P. Suppose $g|_M$ and $g|_P$ are of constant rank. We say the pair (M,P) satisfies <u>condition a_g</u> at p if the following holds:

Let m_i be any sequence of points in M converging to p. Suppose that the sequence of planes $\operatorname{per}(d(g|_M)_{m_i}) \subseteq TA'_{m_i}$ converges to a plane $\tau \subseteq TA'_{m_i}$ in the appropriate Grassmannian bundle. Then $\operatorname{per}(d(g|_M)_p) \subset \tau$.

We say that the pair (M,P) <u>satisfies condition</u> a_g if it satisfies condition a_g at every point p of P.

In order to understand this condition a_g better, let us look at the following example, which should give us some intuitive feeling about this condition.

Let A' = circular cylinder, P = circle as shown in Figure 4.3 and $M = A' - P$. Let A be a cone. Then $g(P)$ = vertex in A and

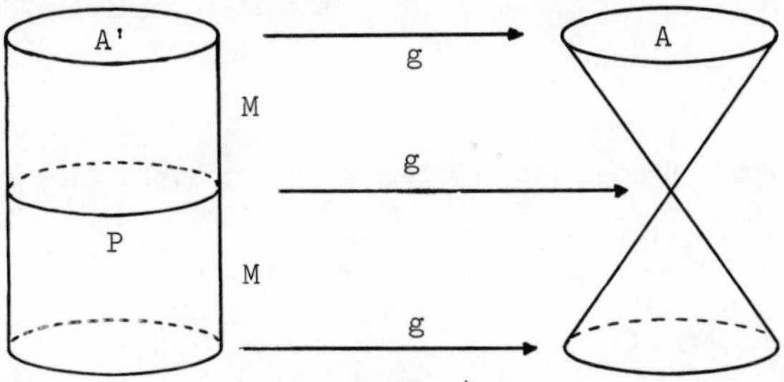

Figure 4.3.

$g(M) = A$ - vertex. In this situation, condition a_g is not satisfied. We would explain this in the following way. As shown in Figure 4.4, let $\{m_i\}$

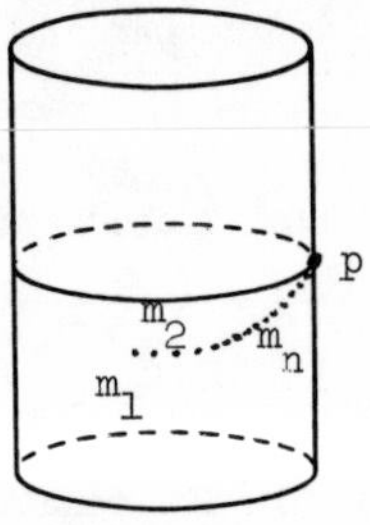

Figure 4.4.

be a sequence of points in M such that $m_i \to p$ as $i \to \infty$. Then

(1) $d(g|_M)m_i$ is an isomorphism of TM_{m_i} with $T(g(M))_{g(m_i)}$

(2) it is clear that $\ker(d(g|_M)_{m_i}) = 0$

(3) since $d(g|P)_p$ is the zero map, $\ker(d(g|P)_p) = (TP)_p$ has dimension one.

Thus $\ker(d(g|P)_p) \not\subset \tau \equiv \lim_{i \to \infty} \ker(d(g|M)_{m_i})$.

Now we return to the situation of Diagram 4.4. We will say that g is a <u>Thom mapping</u> (over $\mathbb{R}$) if the following conditions are satisfied.

(a) $g|_{V'}$ and $f|_V$ are proper.

(b) For each stratum M of V, $f|_M$ is a submersion.

(c) For each stratum M' of S', $g(M')$ lies in a stratum M of V, and $g: M' \to M$ is a submersion (whence $g|_{M'}$ is of constant rank).

(d) Any pair (M',P') of strata of V' satisfies condition a_g (which makes sense in view of (c)).

Theorem 4.6. (Thom's Second Isotopy Lemma) If g is a Thom mapping over $\mathbb{R}$, then g is locally trivial over $\mathbb{R}$.

For the proof, see [47].

5. Ratio Test

There arises the following natural question: given a stratification $V = M_1 \cup \cdots \cup M_s$, how can we decide whether the stratification is (a)-(b)-regular?

Usually condition (a) can be easily checked; however, (b') is not usually so easy to check.

Kuo [32] has provided a sufficient condition (by analytic methods), called the ratio test, for the (a)-,(b)-regularity conditions to be satisfied. In case the lower dimensional stratum is of dimension one then this condition is also necessary.

Since complex varieties in $\mathbb{C}^n$ can be considered as real varieties in $\mathbb{R}^{2n}$, we shall only consider the real case. Further since the regularity conditions are of a local nature we can therefore restrict ourselves to the following situation in $\mathbb{R}^n$: Let $P \subset \overline{M}$ where P, M are two local semi-analytic sets in $\mathbb{R}^n$. Assume that P is an analytic manifold containing $\overline{0} \in \overline{M} - M$, and that P coincide with $T(P,\overline{0})$ at $\overline{0}$. (If not, we can achieve this by an analytic transformation.)

Theorem 5.1. (Ratio Test) For M, P as above, then, if, for every $\tau \in T(P,0)$,

$$\lim_{m \to 0} \frac{|\pi_{N(M,m)}(\tau)|\ |\overline{m}|}{|\overline{m} - \pi_p(\overline{m})|} = 0\,, \qquad (5.1)$$

it follows that M is (a)-(b)-regular over P at $\bar{0}$.

Theorem 5.2. If P is one-dimensional, then (5.1) is a necessary and sufficient condition for M to be Whitney regular over P at $\bar{0}$.

We refer the reader to [32] for the details of the proofs of these two theorems; we will discuss the geometrical meaning of (5.1).

First of all, let us rewrite (5.1) in the following form:

$$\lim_{m \to 0} \frac{|\pi_{N(M,m)}(\tau)|}{\dfrac{|\bar{m} - \pi_p(\bar{m})|}{|\bar{m}|}} = 0 , \qquad (5.2)$$

where the numerator is the measure of the distance between $T(M,m)$ and $T(P,\bar{0})$, as mentioned before, while the denominator can be interpreted as the distance from the unit vector $\frac{\bar{m}}{|\bar{m}|}$ to $T(P,\bar{0})$; see Figure 5.1, where

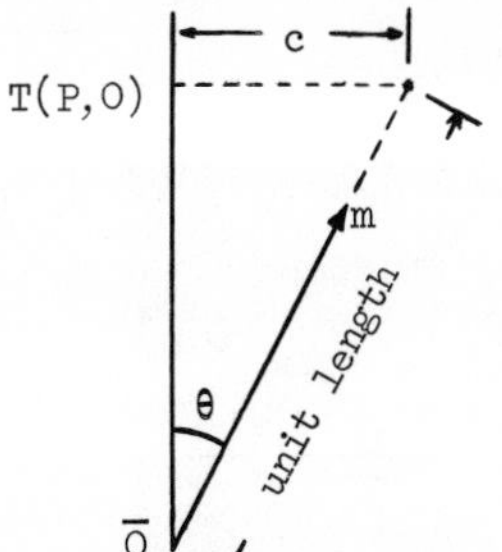

Figure 5.1.

$\frac{|\bar{m} - \pi_p(\bar{m})|}{|\bar{m}|} = \sin\theta = c$, the distance between the unit vector $\frac{\bar{m}}{|\bar{m}|}$ and $T(P,\bar{0})$. Thus (5.1), or (5.2), asserts that, for any $\tau \in T(P,\bar{0})$, the ratio of the above mentioned two quantities tends to zero as $m \to 0$. In other words the distance between $T(M,m)$ and $T(P,\bar{0})$ goes to zero faster than c goes to zero.

Example 5.1. (See Figure 2.4 and Example 2.4.)

$$V = \{(x,y,t) | \; f(x,y,t) = y^2 - t^2x^3 - x^5 = 0\} . \qquad (5.3)$$

Let P = t-axis , M = complement of P in V. At any point $m = (x,y,t)$ with $(x,y) \neq (0,0)$, the normal space is spanned by

$$\text{grad } f = (-3t^2x^2 - 5x^4,\ 2y,\ -2tx^3)\ . \tag{5.4}$$

The normal vector is

$$\nu(m) = \frac{\text{grad } f}{|\text{grad } f|}\ ,$$

and $\tau = (0,0,1)$. Then

$$\begin{aligned} |\pi_{N(M,m)}(\tau)| &= |\nu(m) \cdot \tau| \\ &= \frac{|2tx^3|}{\sqrt{(3t^2x^2 + 5x^4)^2 + 4x^3(t^2 + x^2) + 4t^2x^6}} \end{aligned} \tag{5.5}$$

which approaches zero as $x \to 0$ and $t \to 0$. Since $\pi_p(\bar{m}) = (x,y)$, we have

$$\frac{|\pi_{N(M,m)}(\tau)|}{\dfrac{|\bar{m} - \pi_p(\bar{m})|}{|\bar{m}|}} = \frac{|\nu \cdot \tau|}{\dfrac{\sqrt{x^2 + y^2}}{\sqrt{x^2 + y^2 + t^2}}} = \frac{|\nu \cdot \tau|\sqrt{x^2 + y^2 + t^2}}{\sqrt{x^2 + y^2}} \tag{5.6}$$

$$\leq |\nu \cdot \tau| + \frac{|\nu \cdot \tau|\ |t|}{\sqrt{x^2 + y^2}}\ .$$

From (5.5) we observed that $\lim_{m \to 0} |\nu \cdot \tau| = 0$. Now we claim that

$$\lim_{m \to 0} \frac{|\nu \cdot \tau|\ |t|}{\sqrt{x^2 + y^2}} = 0\ . \tag{5.7}$$

In fact,

$$\frac{|\nu \cdot \tau|\ |t|}{\sqrt{x^2 + y^2}} = \frac{|\nu \cdot \tau|\ |t|}{\sqrt{x^2 + t^2x^3 + x^5}} \tag{5.8}$$

For $|m|$ small, $\sqrt{x^2 + t^2x^3 + x^5} \geq \frac{1}{2}|x|$ since t^2x^3 and x^5 are higher order terms. Thus

$$\lim_{m \to 0} \frac{|\nu \cdot \tau|\ |t|}{\sqrt{x^2 + y^2}} \leq \lim_{m \to 0} \frac{2|\nu \cdot \tau|\ |t|}{|x|} . \tag{5.9}$$

From (5.5) again,

$$\lim_{m \to 0} \frac{|\nu \cdot \tau|\ |t|}{\sqrt{x^2 + y^2}} \leq \lim_{m \to 0} \frac{4|t\,x^3|}{|x|\ 2|x|^{3/2}(t^2 + x^2)^{1/2}} = 0 \tag{5.10}$$

This proved that condition (5.2) holds and hence, by Theorem 5.1, M is (a)-,(b)-regular over P.

<u>Remark</u>. If we were dealing with the <u>complex</u> case, then $t^2 + x^2$ could go to zero and the stratification would not even be (a)-regular.

<u>Example 5.2</u>. The slow spiral (see Figure 2.5 and Example 2.5)

$$V = \{(r,\theta) \mid r = \begin{cases} e^{-\theta} \\ 0 & \text{as } \theta \to \infty \end{cases} \} .$$

Let $P = \{0\}$ and M = complement of P in V. It is easy to check that in this case the ratio test (5.1) is satisfied but the stratification is not (b')-regular. Thus the ratio test fails in this case. (Of course, P is not one-dimensional.)

<u>Example 5.3</u>. The quick spiral (see Figure 2.6). The stratification as Example 5.2 does satisfy the Whitney regularity conditions.

<u>Final Remark</u>. From the last two examples, it is quite clear to the reader that we are seeing some kind of unsatisfactory situation. The task then is to find a regularity condition (c) which will get rid of the unsatisfactory aspects of the situation.

CHAPTER 6

C^0-SUFFICIENCY OF JETS

1. Introduction

The notion of universal unfolding has been discussed in this book on many occasions. The question of dimensional minimality of the universal unfolding will be one of our concerns in this chapter. Recall that in the definition of the universal unfolding, the map Φ is a C^∞-mapping. Thus, according to Theorem 2.2 of Chapter 3, the universal unfolding of the germ $x^4 + y^4$ is of codimension eight. However, it seems that if the map Φ is only required to be continuous (C^0-) then the codimension of the universal unfolding of the germ $x^4 + y^4$ is seven [18]. This motivates us to discuss C^0-sufficiency in greater depth.

Now, the most fundamental problem in local differential analysis is to determine the local topological behavior of a mapping $f: \mathbb{R}^n \to \mathbb{R}^p$, with $f(\bar{0}) = \bar{0}$, near $\bar{0} \in \mathbb{R}^n$, as well as the local topological picture of the variety $f^{-1}(\bar{0})$. Just as in Chapter 2, two such mappings f and g are said to have the same local topological type (or, to be locally topological equivalent or C^0-equivalent) if there exist local homeomorphisms $h_1: \mathbb{R}^n \to \mathbb{R}^n$, $h_1(\bar{0}) = \bar{0}$ and $h_2: \mathbb{R}^p \to \mathbb{R}^p$, $h_2(\bar{0}) = \bar{0}$, such that the diagram

$$\begin{array}{ccc} \mathbb{R}^n & \xrightarrow{\ f\ } & \mathbb{R}^p \\ {\scriptstyle h_1}\downarrow & & \downarrow{\scriptstyle h_2} \\ \mathbb{R}^n & \xrightarrow[\ g\ ]{} & \mathbb{R}^p \end{array}$$

is locally commutative. f and g are said to have the same local topological picture if their varieties at zero, $f^{-1}(\bar{0})$ and $g^{-1}(\bar{0})$, are homeomorphic. Of course, the local topological type of a mapping determines its local topological pictures. But the converse is in general not true. However, we shall find in section 2 of this chapter (see also [28]), that for a real-valued function f

the converse is true when $\bar{0}$ is a topologically isolated singularity of f, i.e. when, for some $r > 0$ and $\epsilon > 0$,

$$|\text{grad } f(x)| \geq \epsilon |x|^r$$

for all x near $\bar{0} \in \mathbb{R}^n$.

If, in the above definition, one replaces the homeomorphisms h_1 and h_2 by C^m-diffeomorphisms $(m \leq \infty)$, then one speaks of the C^m-equivalence of mappings of f and g.

The problem of determining the local behavior of mappings and of varieties is an old and familiar one as we indicated in Chapter 2. For example, if at $\bar{0} \in \mathbb{R}^n$ the Jacobian matrix of f has maximal rank, then the Implicit Function Theorem (Theorem 3.5, Chapter 1) implies that the local behavior of f is completely determined by the knowledge of the first order partial derivatives of f, or $j^{(1)}(f)$, and $f^{-1}(0)$ is locally a differentiably embedded submanifold. In particular, if the first order partial derivatives of a mapping $f: \mathbb{R}^n \to \mathbb{R}$, $f(\bar{0}) = 0$, do not all vanish at $\bar{0} \in \mathbb{R}^n$, then the linear terms of its Taylor series expansion about $\bar{0}$ determines the local behavior of f; and in this case $f^{-1}(0)$ is locally a hyperplane.

The situation becomes more complicated when the Jacobian matrix does not attain maximal rank at $\bar{0} \in \mathbb{R}^n$ for a mapping $f: \mathbb{R}^n \to \mathbb{R}^1$, $f(\bar{0}) = 0$. The Implicit Function Theorem does not apply to the singular point $\bar{0}$. However, if f is real-valued and has a non-vanishing Hessian, i.e.

$$\left|\left(\frac{\partial^2 f}{\partial x_i \partial x_j}\right)\right| \neq \bar{0}$$

at $0 \in \mathbb{R}^n$, then by Marston Morse's Theorem [51] the knowledge of the second order partial derivatives is sufficient for determining the local behavior of f and $f^{-1}(0)$ is locally a quadratic cone.

On the other hand, there are functions for which the knowledge of all of the partial derivatives of any order is not sufficient for determining the local topological picture of their varieties. The flat function $\exp(-\frac{1}{x^2})$ at $x = 0$, is one such example.

Motivated by the above discussion, let us formulate our problems as follows: let $f = (f_1,\ldots,f_p): \mathbb{R}^n \to \mathbb{R}^p$ with $f(\bar{0}) = \bar{0}$ be a local C^s-mappings. Expanding each f_i into a formal Taylor series expansion about the origin, we may write, for any integer r with $1 \leq r \leq s$,

$$f(x) = H_1(x) + H_2(x) + \ldots + H_r(x) + \text{remainder}\,,$$

where $x = (x_1,\ldots,x_n)$ and each $H_i(x)$ is a p-tuple of homogeneous polynomials of degree i, $i = 1,\ldots,r$.

<u>Problem 1</u>. Find an integer r such that the local behavior of f is determined by $H_1 + \ldots + H_r$, a p-tuple of polynomials of degree r. Or, at least, find an r such that the local picture of the variety $f^{-1}(\bar{0})$ is determined by

$$(H_1 + \ldots + H_r)^{-1}(\bar{0})\,.$$

<u>Problem 2</u>. Find the smallest such r.

Besides those classical results stated earlier, not much progress had been made in these problems until about a decade ago. Especially in the last several years many important results in this area have been obtained. Without any doubt these developments will have a profound effect on many fields of mathematics as well as on some applied sciences. It is the purpose of this chapter to summarize these results with illustrative examples and give extensions to some theorems. Also, in the last section we shall point out some interesting open questions in this newly developed field.

To study the problems mentioned above, it is clear that the notion of jets (see Chapter 2) is very convenient.

<u>Definition 1.1</u>. An r-jet $Z \in J^r(n,p)$ is said to be <u>C^0-sufficient</u> in C^{r+1} if any two C^{r+1}-mappings f and g, both realizing Z, have the same local topological behavior. If f and g are locally C^m-equivalent with $0 \leq m \leq r + 1$,

then we call Z <u>C^m-sufficient</u> in C^{r+1}. And Z is said to be <u>v-sufficient</u> in C^{r+1} (where "v" stands for "variety") if the varieties $f^{-1}(\bar{0})$ and $g^{-1}(\bar{0})$ are locally homeomorphic.

Clearly C^0-sufficiency implies v-sufficiency. From the above definition we see that if an r-jet Z is C^0-sufficient (v-sufficient, respectively) in C^{r+1} then adding to Z any terms (perturbation) of degree $> r$ will not change the local topological behavior of the polynomial Z (the local topological picture of $Z^{-1}(\bar{0})$, respectively). Or put it another way, if for a given function f one can find an integer r such that $j^{(r)}(f)$ is sufficient, then as far as local properties are concerned one need only study the truncated polynomial $j^{(r)}(f)$. Before we provide examples, lèt us define one more notion.

<u>Definition 1.2</u>. For a local C^s-mapping $f: \mathbb{R}^n \to \mathbb{R}^p$, $f(\bar{0}) = \bar{0}$, the smallest integer k with $k < s$ such that $j^{(k)}(f)$ is C^0-sufficient (C^m-sufficient, v-sufficient, respectively) in C^{k+1} is called the <u>degree of C^0-sufficiency</u> (C^m-sufficiency, v-sufficiency, respectively) of f in C^{k+1}.

Some of the examples provided in the following are the same as the examples given in section 1 of Chapter 3. We encourage the reader to compare their roles in these two sections as this is interesting in itself.

<u>Example 1.1</u>. For any positive integer r, the jet in $J^r(2,1)$ given by $Z(x,y) = x^2$ is not a C^0-sufficient r-jet. For there is an integer N such that $2N > r$, and the variety of the realization $f(x,y) = x^2 - y^{2N}$ and the variety of Z are clearly not homeomorphic, since $f^{-1}(0)$ has two branches at 0 and $Z^{-1}(0)$ is just a line. Thus Z is not v-sufficient, hence not C^0-sufficient and obviously not C^∞-sufficient.

<u>Remark</u>. From the above example we see that any polynomial of the form $x^2 g(x,y)$ is not C^0-sufficient as an r-jet for any r (so long as $g(x,y)$ is a C^{r+1}-function), since the realization $(x^2 - y^{2N})g(x,y)$ is not v-equivalent to $x^2 g(x,y)$ where $2N > r$. In fact, this is also a special case of Corollary 4.3 of this chapter.

Example 1.2. The 3-jet $Z \in J^3(2,1)$ given by $Z(x,y) = x^2 + y^3$ is C^0-sufficient in C^4. That is, any perturbation of degree > 4 will not change its local topological behavior. The argument is the same as in section 1 of Chapter 3. Therefore, the degree of C^0-sufficiency in this jet is three.

Example 1.3. Consider $Z(x,y) = x^2 - xy^3 \in J^6(2,1)$, i.e. as a 6-jet. By the method to be discussed in section 2 it is easy to show that Z is a C^0-sufficient 6-jet in C^7, i.e. the local topological behavior is unchanged with perturbations of degree ≥ 7. However, if one considers Z as a 5-jet, then it is not C^0-sufficient in C^6. For

$$f(x,y) = x^2 - xy^3 - \frac{3}{4}y^6 = (x - \frac{1}{2}y^3)^2 - y^6 ,$$

and

$$g(x,y) = x^2 - xy^3 + \frac{5}{4}y^6 = (x - \frac{1}{2}y^3)^2 + y^6$$

are C^6-realizations of the 5-jet Z, i.e.

$$j^{(5)}(f) = j^{(5)}(g) = Z ,$$

but $f^{-1}(0)$ and $g^{-1}(0)$ have different local topological pictures. Thus $j^{(6)}(Z)$ is C^0-sufficient in C^7, while $j^{(5)}(Z)$ is not C^0-sufficient in C^6. Thus we see that the degree of C^0-sufficiency of Z is 6 (in C^7).

The problem of sufficiency can also be formulated with respect to other classes of perturbations. For instance, we call an r-jet $Z \in J^r(n,p)$ <u>C^0-sufficient in C^r</u> if any two C^r-realizations of Z have the same local topological behavior. For a local C^s-mapping f, the smallest integer k with $k \leq s$ such that $j^{(k)}(f)$ is C^0-sufficient in C^k is called <u>the degree of strong C^0-sufficiency of</u> f. One can formulate similar definitions for v- and C^m-sufficiency. In this chapter, we shall state results for the sufficiency of r-jets mostly in the sense of Definitions 1.1, 1.2. With simple modifications as pointed out in section 2, most of the results will hold also for the strong sufficiency of r-jets. However, there is a difference between these two notions of sufficiency as one can see in the next example. For the discussion of C^0- and v-sufficiency of an r-jet in C^∞ and in C^ω when $p = 1$, where C^ω is the set of all convergent power series, we refer the reader to Bochnak and Łojasiewicz [8].

Example 1.4. Consider $Z(x,y) = x^3 - 3xy^7 \in J^{10}(2,1)$. We will see in section 2 (Example 3.1) that the degree of C^0-sufficiency of Z is 10 (with C^{11} perturbations). However, this 10-jet Z is not C^0-sufficient with C^{10} perturbations, i.e. the strong degree of C^0-sufficiency is not even defined. For the 10-jet Z has the following C^{10}-realization

$$f(x,y) = x^3 - 3xy^7 + 2|y|^{10\frac{1}{2}} = (x - |y|^{\frac{7}{2}})^2(x + 2|y|^{\frac{7}{2}}),$$

and perturbations of the form, $\pm y^{2N}(x + 2|y|^{\frac{7}{2}})$ with large N, to $f(x,y)$ will give rise to non-homeomorphic varieties near $\bar{0} \in R^2$.

Our main concern in the remaining sections is to determine C^0- and v-sufficiency of jets. In section 2 we survey the results in $J^r(n,1)$ and in section 3 we describe a step-by-step method for finding the degree of C^0-sufficiency of jets in $J^r(2,1)$. Then we extend Theorem 3.2, which is about decomposable jets in $J^r(2,1)$, to one about decomposable jets in $J^r(n,1)$ (Theorem 3.3). In section 4 we summarize the results on v-sufficiency in $J^r(n,p)$ and also give a sufficient condition (Theorem 3.6) for an r-jet in $J^r(n,p)$ to be v-sufficient under analytic perturbation, i.e. v-sufficient in C^ω.

2. Criterion of C^0- and v-Sufficiency of Jets in $J^r(n,1)$

The following is the fundamental theorem which characterizes the C^0- and v-sufficiency of r-jets in $J^r(n,1)$ with C^{r+1} perturbation (a C^{r+1} perturbation is a C^{r+1} function P such that $j^{(r)}(P) = 0$). For complex cases, the criterion [38] is the same as this as this theorem.

Theorem 2.1. Let Z be an r-jet in $J^r(n,1)$. Then the following conditions are equivalent:

(a) Z is C^0-sufficient in C^{r+1},

(b) Z is v-sufficient in C^{r+1},

(c) there exist constants $\epsilon > 0$ and $\delta > 0$ such that

$$|\text{grad } Z(x)| \geq \epsilon |x|^{r-\delta},$$

for all x in a neighborhood of $\bar{0} \in \mathbb{R}^n$.

The implication (c) ⇒ (a) was first discovered by N. H. Kuiper [27] for $\delta = 1$, and was later established independently by T. C. Kuo [28] whose proof is valid for any $\delta > 0$. Kuo's proof will be presented since it is simple, articulate and most importantly, the technique in the proof will be used to prove that the codimension of the topological universal unfolding of $x^4 + y^4$ is seven. The implication (b) ⇒ (c) was first conjectured by R. Thom and later proved by Bochnak and Łojasiewicz. We refer [8] to the reader for this proof. The implication (a) ⇒ (b) is of course trivial. Thus, in $J^r(n,1)$, C^0- and v-sufficiency of r-jets in C^{r+1} are equivalent notions. The essential ingredient in this theorem is clearly the gradient condition (c). Let us first look at it by means of an example.

<u>Example 2.1</u>. Let $Z(x,y) = x^2 + y^3$. Then

$$|\text{grad } Z(x,y)| = (4x^2 + 9y^4)^{1/2} \geq 2(x^2 + y^2) = 2|(x,y)|^{3-1}$$

for all (x,y) near $0 \in \mathbb{R}^2$, and hence Z is a C^0-sufficient 3-jet in C^4. Indeed, this gives us the same result as in Example 1.2 of this chapter. As a matter of fact, we will learn from Theorem 2.8 that Z is also a C^0-sufficient 3-jet in C^3.

For the proof of the implication (c) ⇒ (a), let us write $Z(x) = H_1(x) + \dots + H_r(x)$, as a truncated polynomial, where $H_i(x)$ is a homogeneous form of degree i. If $H_1(x) \neq 0$, the Implicit Function Theorem implies that Z is C^0-sufficient. Thus, we assume $H_1 \equiv 0$.

Let $P(x)$ be a C^{r+1} perturbation, then $\lim\limits_{x \to 0} \dfrac{|P(x)|}{|x|^{r+1-\delta}} = 0$ and obviously $\lim_{x \to 0} \frac{|\text{grad } P(x)|}{|x|^{r-\delta}} = 0$.

Let $F(x,t) = Z(x) + tP(x)$ where $t \in \mathbb{R}$. It is clear that for any t, $\text{grad } F = (\frac{\partial F}{\partial x_1}(x), \ldots, \frac{\partial F}{\partial x_n}(x), P(x)) = 0$ when $x = \bar{0}$. Hence the level surfaces $F = \text{constant}$ have $(\bar{0},t)$, or t-axis, as a line of singularities. We make a further observation: if we can find a local homeomorphism $h: \mathbb{R}^n \to \mathbb{R}^n$ such that $F(x,0) = F(h(x),1)$, then $Z(x) = Z(h(x)) + P(h(x))$ which, by definition, means that Z is C^0-sufficient. In order to do so, we are going to construct a continuous vector field to flow (intuitively) from the hyperplane $t = 0$ to the hyperplane $t = 1$. For this purpose, we construct the vector field as follows. Let $X(x,t)$ be the projection of the vector $(0,1) \in \mathbb{R}^n \times \mathbb{R}$ to the direction of grad F at (x,t), $0 < |x| < \alpha$ where α is small. Then

$$X(x,t) = \frac{|(\bar{0},1) \text{ grad } F|}{|(\bar{0},1)|\,|\text{grad } F|} \frac{\text{grad } F}{|\text{grad } F|} = \frac{|P(x)|\text{grad } F}{|\text{grad } F|^2}.$$

Let

$$Y(x,t) = \begin{cases} (\bar{0},1) - X(x,t) & \text{for } 0 < |x| < \alpha \\ (\bar{0},1) & \text{for } x = \bar{0} \end{cases}$$

<u>Remark</u>. Grad F is in the normal direction of the level surface of $F = \text{constant}$ at each (x,t). Hence Y is tangent to the level surface $F = \text{constant}$ at each (x,t).

Now, before we conclude that the solution curves of the vector field Y will provide the required flow, let us carry out some calculations.

<u>Lemma 2.2</u>. Let Z satisfy condition (c) in Theorem 2.1, and let P be a C^{r+1} perturbation. Then there exists $\alpha > 0$ such that $|\text{grad } F| \geq \frac{\epsilon}{2}|x|^{r-\delta}$ for $t \in [0,1]$, where ϵ and δ were also given in (c) of Theorem 2.1.

<u>Proof</u>: Since $\lim_{x \to 0} \frac{|\text{grad } P(x)|}{|x|^{r-\delta}} = 0$, for any $\epsilon > 0$, we can choose α

small enough such that $\frac{|\text{grad } P|}{|x|^{r-\delta}} < \frac{\varepsilon}{2}$ if $0 < |x| < \alpha$. For $t \in [0,1]$, $x \neq 0$,

$$\begin{aligned}|\text{grad } F| &= |\text{grad}(Z + tP)| \\ &\geq |\text{grad } Z + t \text{ grad } P| \\ &\geq |\text{grad } Z| - |\text{grad } P| \\ &\geq |x|^{r-\delta}\{\varepsilon - \frac{|\text{grad } P|}{|x|^{r-\delta}}\} \\ &\geq \frac{\varepsilon}{2}|x|^{r-\delta} .\end{aligned}$$

Lemma 2.3. $Y(x,t)$ is C^0 for $0 \leq |x| < \alpha$, C^r for $0 < |x| < \alpha$ and

$\lim\limits_{x \to 0} \frac{|Y(x,t) - Y(\bar{0},t)|}{|x|} = 0$ uniformly for $t \in [0,1]$.

Proof: $\lim\limits_{x \to \bar{0}} \frac{|Y(x,t) - Y(\bar{0},t)|}{|x|} = \lim\limits_{x \to \bar{0}} \frac{|X(x,t)|}{|x|} = \lim\limits_{x \to \bar{0}} \frac{|P(x)|}{|x|\ |\text{grad } F|}$.

(i) For $x \neq \bar{0}$, by Lemma 2.2, grad $F \neq 0$, then Y is C^r in $0 < x < \alpha$.

(ii) As $x \to 0$, $\lim\limits_{x \to \bar{0}} \frac{|P(x)|}{|x|\ |\text{grad } F|} \leq \lim\limits_{x \to \bar{0}} \frac{2|P(x)|}{\varepsilon|x|\,|x|^{r-\delta}} = \lim\limits_{x \to \bar{0}} \frac{2|P(x)|}{\varepsilon|x|^{r+1-\delta}} = 0$

for $0 \leq |x| < \alpha$, where α is small. So, we are done.

Note that our purpose is to construct a flow, in other words, a unique local solution of $Y(x,t)$ for each initial point, such that the terminal value is a continuous function of the initial value and the time coordinate u. An existance and uniqueness theorem in the theory of ordinary differential equation is needed since Lemma 2.3 does not provide us with the usual Lipschitz condition. However, from Picard's existence theorem in the theory of ordinary differential equations, there is a local solution of $Y(x,t)$, say $\varphi(x,t;\ u)$ with $\varphi(x_0,t_0;\ 0) = (x_0,t_0)$ where $|u| < \eta$ for some positive real number η.

Definition 2.1. $Y(x,t)$ satisfies the Lipschitz condition along φ with Lipschitz constant $K = K(\varphi)$ if for any (x,t) in a neighborhood of the point set $\{\varphi(x_0,t_0;\ u) \mid |u| < \eta\}$, $|Y(x,t) - Y(\varphi(x_0,t_0;\ u))| < K|(x,t) - \varphi(x_0,t_0;\ u)|$.

Lemma 2.4. Let $Y(x,t)$ be defined as above mentioned. For each (x_0,t_0) in a neighborhood $N = \{(x_0,t_0) | \; 0 \leq |x_0| < \alpha, \; t_0 \in [0,1]\}$, let $\varphi(x_0,t_0; u)$ be a solution with $\varphi(x_0,t_0; 0) = (x_0,t_0)$ along which Y satisfies the Lipschitz condition. Then Y admits a unique solution through each point of N and furhter the solution $\varphi(x_0,t_0; u)$, defined on its maximum interval of existence, is a continuous function of u and the initial point (x_0,t_0).

For the proof of this lemma, we refer the reader to Kuo's paper [28].

It is clear that Lemmas 2.2 and 2.3 imply that $Y(x,t)$ does satisfy the hypothesis of Lemma 2.4. Thus Y admits a unique solution, say $\varphi(x_0,t_0; u)$ with $\varphi(x_0,t_0; u_0) = (x_0,t_0)$, and the terminal value is a continuous function of the initial value as well as the time variable u. The solution $\varphi(\overline{0},0; u) = (\overline{0},u)$ obviously satisfies the initial condition $\varphi(\overline{0},0; 0) = (\overline{0},0)$.

Now the question is whether the solution φ will stop at some t, where for $0 < t < 1$, that is, before φ reaches to $t = 1$. In other words, we must ask whether the solution curve will hit the singularities (or t-axis) of the level surface of $F = \text{constant}$. In order to ensure us this will not happen, we prove the following lemma:

Lemma 2.5. There exists a positive real number $\beta > 0$ so that for $0 \leq |x| < \beta$, the inner product $(Y(x,t),(0,1)) > 0$.

Proof: $(Y(x,t),(0,1)) = 1 - \dfrac{P(x)^2}{|\text{grad } F|^2} \geq 1 - \dfrac{4P(x)^2}{\epsilon^2 |x|^{2r-2\delta}} > 0$ for $|x|$ small.

Proof of Theorem 2.1. For all x in a neighborhood of $\overline{0} \in \mathbb{R}^n$, by Lemma 2.5, the t-component of any solution $\varphi(x,0; u)$ increases monotonically with u since φ is the integral curve of Y. Hence $\varphi(x,0; t)$ meets the hyperplane $t = 1$ at a unique point $h(x)$. The mapping $x \to h(x)$ is a local homeomorphism. Finally, by the remark above, we know that Y is tangent to the level surface $F = \text{constant}$ at each point (x,t), so that F is constant along each φ. Hence

$$Z(x) = F(x,0) = F(h(x),1) = Z(x) + P(x).$$

It is known, by using Mather's results [43], that the codimension of the universal unfolding of $x^4 + y^4$ is eight. In fact, the unfolding of $x^4 + y^4$ is the double cusp polynomial,

$$x^4 + y^4 + tx^2y^2 + ax^2y + bxy^2 + cx^2 + dxy + ey^2 + us + vy ,$$

which is defined by Godwin [18]. We will prove in the following corollary that $x^4 + y^4$ and $x^4 + y^4 + tx^2y^2$ are C^0-equivalent in case $t > -2$, thus the codimension of the topological universal unfolding is seven.

<u>Corollary 2.6</u>. If $t > -2$, $x^4 + y^4$ and $x^4 + y^4 + tx^2y^2$ are C^0-equivalent.

<u>Proof</u>: Let $Z(x,y) = x^4 + y^4$ and $F(x,y,t) = x^4 + y^4 + tx^2y^2$. So long as we can prove that $|\text{grad } F| \geq \varepsilon|(x,y)|^{4-1}$ for some small ε and $t > -2$, the remainder of the proof of this corollary is the same as that of Theorem 2.1 and may be left as an exercise for the reader.

Now

$$\begin{aligned}|\text{grad } F| &= |(4x^3 + 2txy^2,\ 4y^3 + 2tx^2y,\ x^2y^2)| \\ &\geq [4x^2(2x^2 + ty^2)^2 + 4y^2(2y^2 + tx^2)^2]^{1/2} \\ &= (x^2 + y^2)^{1/2}[16(x^2 + y^2)^2 + 4(t^2 + 4t - 12)x^2y^2]^{1/2} .\end{aligned}$$

Observe that $(\frac{x}{y} + \frac{y}{x})^2 \geq 4$ and that in case $t > -2$, there is ε such that $(2 + t)^2 > \varepsilon^2$. Thus $t > -2$ implies that there is $\varepsilon > 0$ such that

$$1 > \frac{12 - 4t - t^2}{16 - \varepsilon^2} .$$

With these 2 observations, it is clear that

$$(\frac{x}{y} + \frac{y}{x})^2 \geq \frac{4(12 - 4t - t^2)}{16 - \varepsilon^2} ,$$

equivalently,

$$[16(x^2 + y^2)^2 + 4(t^2 + 4t - 12)x^2y^2]^{1/2} \geq \varepsilon(x^2 + y^2) .$$

Hence, we have

$$|\text{grad } F| \geq \epsilon |(x,y)|^3 .$$

It is not difficult to see that the gradient condition (c), for $\delta = 1$, is sufficient to show that the r-jet is C^0-sufficient in C^r. (The exercise left in the proof of Corollary 2.6 precisely serves this purpose.) We obtain:

Theorem 2.7 [8]. Let Z be an r-jet in $J^r(n,1)$. Then the following conditions are equivalent:

(a) Z is C^0-sufficient in C^r,

(b) Z is v-sufficient in C^r,

(c) there exists a constant $\epsilon > 0$ such that

$$|\text{grad } Z(x)| \geq \epsilon |x|^{r-1} ,$$

for all x in a neighborhood of $\overline{0} \in \mathbb{R}^n$.

Thus, the C^0-sufficiency of r-jets can be formulated in C^r. It is very important to point out that the notion of C^0-sufficiency of an r-jet in C^{r+1} and in C^r are quite different as, one can see in Example 1.4.

On the other hand, since we mentioned C^k-sufficiency in section 1, it is appropriate to consider the following corollary and example. The technique of the proof of Theorem 2.1 can also be used to establish the following corollary:

Corollary 2.8. Let $H_r(x)$ be a homogeneous polynomial of degree r. Then H_r is a C^1-sufficient jet in C^{r+1} if there exists $\epsilon > 0$ and $\delta > 0$ such that

$$|\text{grad } H_r(x)| \geq \epsilon |x|^{r-\delta}$$

for all x in a neighborhood of $\overline{0} \in \mathbb{R}^n$.

When $r = 2$, this implies the well-known Morse Theorem [51]. For arbitrary

r, the above result is the best that one can get with respect to the smoothness of sufficiency as illustrated in the next example.

Example 2.3. Consider $Z(x,y) = x^5 + y^5 \in J^5(2,1)$. Then by Corollary 2.8, Z is a C^1-sufficient 5-jet in C^6. However, Z is not C^2-sufficient in C^6. For Z and the C^6-realization f, given by

$$f(x,y) = x^5 + y^5 + x^3y^3$$

are not C^2-equivalent. This can be seen as follows. Let $h: \mathbb{R}^2 \to \mathbb{R}^2$, $h(0,0) = (0,0)$, be a local C^2-diffeomorphism. Then h can be expressed as $h(x,y) = (h_1(x,y), h_2(x,y))$, where

$$h_1(x,y) = a_1x + b_1y + c_1x^2 + d_1xy + e_1y^2 \quad \text{(modulo terms of degree} > 2\text{)},$$

$$h_2(x,y) = a_2x + b_2y + c_2x^2 + d_2xy + e_2y^2 \quad \text{(modulo terms of degree} > 2\text{)}.$$

But the term x^3y^3 cannot be obtained in

$$Z(h(x,y)) = Z(h_1(x,y), h_2(x,y)) = (h_1(x,y))^5 + (h_2(x,y))^5 .$$

On the other hand, what happens if a jet is not C^0-sufficient as an r-jet for any r, for example $Z(x,y) = x^2$ in $J^r(2,1)$? Thom conjectured that if an r-jet $Z \in J^r(n,p)$ is not C^0-sufficient in C^{r+1} (or in C^r) then Z has an infinite family of realizations $\{f_\alpha\}$ such that, whenever $\alpha \neq \beta$, f_α and f_β have different local topological behavior (see Manifold, Lecture Notes in Math. No. 197, p. 229 Problem 3). The case $n = 2$ and $p = 1$ was first proved by T. C. Kuo [28]. In 1972, Bochnak and T. C. Kuo [7] proved the following theorem which is the case $p = 1$ with an arbitrary n.

Theorem 2.9. If $Z \in J^r(n,1)$ is not C^0-sufficient in C^{r+1} (or in C^r), then there exists an infinite sequence $\{f_i \mid i = 1, 2, \ldots\}$ of realizations of Z such that, whenever $i \neq j$, the varieties $f_i^{-1}(0)$ and $f_j^{-1}(0)$ are not homeomorphic.

How large is the set of mappings which satisfy the equivalent conditions of Theorem 3.1 ? The following theorem, which is a special case of Corollary 4.3, says that this set is very large.

Theorem 2.10. If a jet $Z \in J^{r-1}(n,1)$ is given, then for "almost all" homogeneous polynomials $H_r(x)$ of degree r, we have

$$|\mathrm{grad}(Z(x) + H_r(x))| \geq \varepsilon |x|^{r-1}$$

for all x in a neighborhood of $0 \in R^n$, where $\varepsilon > 0$ is a constant depending on H_r. Hence, as an r-jet, $Z + H_r$ is C^0-sufficient in C^{r+1}.

The set of all homogeneous polynomials H_r constitutes a Euclidean space R^N when the coefficients of H_r are ordered in any fixed manner. By "almost all H_r" we mean all H_r except possibly those in a proper algebraic subvariety of R^N.

3. Degree of C^0-Sufficiency

Now suppose given a polynomial function or a formal power series f and suppose that $j^{(r)}(f)$ is C^0-sufficient for some finite r; how can we find the smallest integer k such that $j^{(k)}(f)$ is C^0-sufficient in C^{k+1}, i.e. how may we find the degree of C^0-sufficiency? A complete answer is available only for the jet space $J^r(2,1)$; this was first done by T. C. Kuo [29] and later improved in [37]. Here we describe in the following a step-by-step method for determining the degree of C^0-sufficiency of a given polynomial or a formal power series $f(x,y)$ of two variables.

Step 1: Let $f(x,y) = H_a(x,y) + H_{a+1}(x,y) + \dots$, with the initial homogeneous polynomial $H_a(x,y)$, a degree a, factored into q factors

$$H_a(x,y) = P_{a_1}^{(1)} \dots P_{a_q}^{(q)},$$

where each $P^{(i)}_{a_i}$ has degree a_i, $i = 1,\dots,q$, and $P^{(1)}_{a_1},\dots,P^{(q)}_{a_q}$ are pairwise relatively prime. Then according to Theorem 1 in [37] one can always find a formal power series $f_i(x,y)$ with initial form $P^{(i)}_{a_i}$ for each $i = 1,\dots,q$ such that

$$f(x,y) = f_1(x,y) \dots f_q(x,y) .$$

Step 2: For each $i = 1,\dots,q$, determine the degree of C^0-sufficiency of f_i. Using a local C^∞-change of coordinates if necessary ([37], Theorem 4), we may consider without loss of generality a formal power series of the form

$$Z(x,y) = x^t + K_{t+1}(x,y) + K_{t+2}(x,y) + \dots , \qquad (3.1)$$

where the homogeneous forms K_j do not have terms involving any power x^i for $i \geq t-1$. Then by applying Puiseux's Theorem and with the help of a Newton polygon [80, p. 97-105], we can decompose $\frac{\partial Z}{\partial x}$ and $\frac{\partial Z}{\partial y}$ into factors as follows:

$$Z_x(x,y) = t(x - p_1(y))(x - p_2(y)) \dots (x - p_{t-1}(y)) ,$$

where each $p_i(y)$ is a fractional power series of y with order

$$O(p_i(y)) > 1 , \quad i = 1,2,\dots,t-1 ,$$

and

$$Z_y(x,y) = h(x,y)a(y)(x - q_1(y)) \dots (x - q_s(y)) ,$$

where each $q_j(y)$ is a fractional power series of y with order

$$O(q_j(y)) > 1 , \quad j = 1,2,\dots,s ,$$

$h(x,y)$ consists of branches of order ≤ 1, and $a(y)$ consists of y only. Note that if some fractional power series $q_j(y) \equiv 0$ we use the convention $O(q_j(y)) = +\infty$. Let $U_i(y)$ be the real part of $p_i(y)$ and $W_j(y)$ the real part of $q_j(y)$. Let

$$m_i = \min\{O(Z_x(U_i(y),y)),\ O(Z_y(U_i(y),y))\},$$
$$n_j = \min\{O(Z_x(W_j(y),y)),\ O(Z_y(W_j(y),y))\}, \tag{3.2}$$

and then let k be the smallest integer such that

$$k > \max_{\substack{i=1,\ldots,t-1 \\ j=1,\ldots,s}} \{m_i, n_j\}.$$

<u>Theorem 3.1</u>. (Kuo [29]) For a formal power series $Z(x,y)$ as in (3.1), the integer k defined above is the degree of C^0-sufficiency of Z (in C^{k+1}).

<u>Example 3.1</u>. $Z(x,y) = x^3 - 3xy^\ell$, $\ell \geq 3$. Here we have

$$Z_x(x,y) = 3(x - y^{\ell/2})(x + y^{\ell/2}),$$
$$Z_y(x,y) = -3\ell\, xy^{\ell-1}.$$

Hence, $m_1 = m_2 = \ell/2 + \ell - 1$ and $n_1 = \ell$. By Theorem 3.1, the smallest integer k such that $k > \ell/2 + \ell - 1$ is the degree of C^0-sufficiency of Z in C^{k+1}. For instance, when $\ell = 7$, $k = 10$.

<u>Example 3.2</u>. $Z(x,y) = x^4 - 4xy^9$. Then we have

$$Z_x(x,y) = 4(x - y^3)\Big(x + (\tfrac{1}{2} + i\tfrac{\sqrt{3}}{2})y^3\Big)\Big(x + (\tfrac{1}{2} - i\tfrac{\sqrt{3}}{2})y^3\Big),$$

$$Z_y(x,y) = -36xy^8,$$

and hence $m_1 = 11$, $m_2 = m_3 = 9$, and $n_1 = 9$. Therefore, the degree of C^0-sufficiency of Z is 12 in C^{13}.

<u>Step 3</u>: Now, having determined the degree of C^0-sufficiency of each $f_i(x,y)$, $i = 1,\ldots,q$, we can get the degree of C^0-sufficiency of $f(x,y)$ by applying Theorem 2 in [37], which is the following:

Theorem 3.2. As in Step 1, let $f(x,y) = f_1(x,y) \dots f_q(x,y)$ and each f_i have initial form $P^{(i)}_{a_i}$ with $P^{(1)}_{a_1}, \dots, P^{(q)}_{a_q}$ pairwise relatively prime. If k_i is the degree of C^0-sufficiency of f_i in C^{k_i+1}, then the degree of C^0-sufficiency of f is given by

$$m = \sum_{i=1}^{q} a_i + \max_{i=1,\dots,q} \{k_i - a_i\}$$

in C^{m+1}.

Example 3.3. $f(x,y) = x^3y^4 - 3xy^9 - 4x^{12}y + 12x^{10}y^6$. Here we can decompose $f(x,y)$ as

$$f(x,y) = (x^3 - 3xy^5)(y^4 - 4yx^9)$$

with $f_1(x,y) = x^3 - 3xy^5$ and $f_2(x,y) = y^4 - 4yx^9$. By Examples 3.1 and 3.2 we see that $k_1 = 7$ and $k_2 = 12$. Hence

$$m = 3 + 4 + \max\{7 - 3, 12 - 4\} = 15$$

is the degree of C^0-sufficiency of f in C^{16}.

Remark. If for a given $f(x,y)$ the above three steps yield $m = +\infty$, then $j^{(r)}(f)$ is not C^0-sufficient for any finite r.

Theorem 3.2 can be generalized to $J^r(n,1)$ as follows and we shall sketch a proof.

Theorem 3.3. Let $Z = Z(x_1,\dots,x_n)$ be a formal power series such that $Z = Z_1 \dots Z_q$. For each $i = 1,\dots,q$, let k_i be the degree of C^0-sufficiency of Z_i in C^{k_i+1} and

$$Z_i = H^i_{a_i} + H^i_{a_i+1} + \dots,$$

where each $H_j^i = H_j^i(x_1,\dots,x_n)$ is a homogeneous form of degree j. Assume that except at $\overline{0} \in R^n$ the varieties $(H^1_{a_1})^{-1}(0),\dots,(H^q_{a_q})^{-1}(0)$ are pairwise disjoint. Then the integer given by

$$m = \sum_{i=1}^{q} a_i + \max_{i=1,\dots,q} \{k_i - a_i\}$$

is the degree of C^0-sufficiency of Z in C^{m+1}.

<u>Lemma 3.4</u>. Let $P = P(x_1,\dots,x_n)$ be any C^{m+1}-function such that $j^{(m)}(P) = 0$, where m is the number in the above theorem. Then there exists C^{k_i+1}-functions $P_i = P_i(x_1,\dots,x_n)$ with $j^{(k_i)}(P_i) = 0$, $i = 1,2,\dots,q$, such that

$$Z + P \sim (Z_1 + P_1) \dots (Z_q + P_q),$$

where "$\sim$" indicates equivalence under a local C^1-diffeomorphism.

<u>Proof of Lemma 3.4</u>: For simplicity we consider the case $q = 2$. The proof of the general case is similar.

Let S^{n-1} be the unit sphere in R^n, and let D_1 and D_2 be the zeros of $H^1_{a_1}$ and $H^2_{a_2}$ on S^{n-1}. Let N_1 and N_2 in S^{n-1} be closed disjoint neighborhoods of D_1 and D_2 respectively. Choose C^∞-functions g_1 and g_2 on S^{n-1} with disjoint supports such that $g_i = 1$ on an open neighborhood of N_i, $i = 1, 2$. Then for any $(x_1,\dots,x_n) \neq (0,\dots,0)$, we define

$$\tilde{P}_1(x_1,\dots,x_n) = P(x_1,\dots,x_n)g_1\left(\frac{x_1}{r},\dots,\frac{x_n}{r}\right),$$

$$\tilde{P}_2(x_1,\dots,x_n) = P(x_1,\dots,x_n)g_2\left(\frac{x_1}{r},\dots,\frac{x_n}{r}\right),$$

where $r = (x_1^2 + \dots + x_n^2)^{1/2}$, and define

$$\tilde{P}_1(0,\dots,0) = \tilde{P}_2(0,\dots,0) = 0.$$

Let

$$P_1 = \begin{cases} \dfrac{\tilde{P}_1}{Z_2}, & \text{if } (x_1,\ldots,x_n) \notin T_2 \\ 0, & \text{if } (x_1,\ldots,x_n) \in T_2, \end{cases}$$

where T_2 is the union of all radii through N_2 and contains the tangent cone of Z_2. Similarly, let

$$P_2 = \begin{cases} \dfrac{\tilde{P}_2}{Z_1}, & \text{if } (x_1,\ldots,x_n) \notin T_1 \\ 0, & \text{if } (x_1,\ldots,x_n) \in T_1, \end{cases}$$

where T_1 is the union of all radii through N_1 and contains the tangent cone of Z_1.

We see that each P_i is a C^{k_i+1}-function, $j^{(k_i)}(P_i) = 0$, $P_1P_2 = 0$, and

$$P_1' = P_1Z_2 \quad \text{and} \quad P_2' = P_2Z_1$$

for all $(x_1,\ldots,x_n)$. Define

$$g(x_1,\ldots,x_n) = g_1\left(\frac{x_1}{r},\ldots,\frac{x_n}{r}\right) + g_2\left(\frac{x_1}{r},\ldots,\frac{x_n}{r}\right),$$

if $(x_1,\ldots,x_n) \neq (0,\ldots,0)$, and $g(0,\ldots,0) = 0$.

Now let

$$F(x_1,\ldots,x_n,t) = Z_1Z_2 + (t(g-1)+1)P, \; t \in R.$$

Then, using a technique similar to the one that has been discussed in section 2 of this chapter, one can easily show that

$$F(x_1,\ldots,x_n,0) \sim F(x_1,\ldots,x_n,1).$$

This completes the proof of the lemma, since

$$F(x_1,\ldots,x_n,0) = Z_1Z_2 + P$$

and

$$\begin{aligned} F(x_1, \ldots, x_n, 1) &= Z_1 Z_2 + g_1 P + g_2 P \\ &= Z_1 Z_2 + P_1' + P_2' \\ &= (Z_1 + P_1)(Z_2 + P_2) . \end{aligned}$$

Proof of Theorem 3.3: Let P be any C^{m+1}-function with $j^{(m)}(P) = 0$. Then by the above lemma we have

$$Z + P \sim (Z_1 + P_1) \ldots (Z_q + P_q) ,$$

for some C^{k_i+1}-functions P_i with $j^{(k_i)}(P_i) = 0$, $i = 1, \ldots, q$. Also, for each i, k_i is the degree of C^0-sufficiency of Z_i. It follows that the varieties $Z^{-1}(0)$ and $(Z + P)^{-1}(0)$ are locally homeomorphic. This shows that $j^{(m)}(Z)$ is v-sufficient and hence C^0-sufficient in C^{m+1}. Finally, by an argument similar to that in [37] (p. 125), one can show that m is the smallest integer such that Z is C^0-sufficient in C^{m+1}.

Example 3.4. $Z(x,y,z) = (x^2 + y^2 + z^5)(x^2 + z^2 + y^3)$. Here we have $Z_1 = x^2 + y^2 + z^5$, $Z_2 = x^2 + z^2 + y^3$, and $a_1 = a_2 = 2$. Also, it is easy to see that $k_1 = 5$ and $k_2 = 3$. Thus

$$m = 2 + 2 + \max\{5 - 2, 3 - 2\} = 7$$

is the degree of C^0-sufficiency of Z in C^8.

4. Sufficiency of Jets in $J^r(n,p)$

We have seen in the previous sections that in $J^r(n,1)$ the notions of v-sufficiency and C^0-sufficiency are equivalent. However, in the general jet space $J^r(n,p)$ with $p > 1$, this is no longer true as one can see from the following example.

Example 4.1. Consider $Z(x,y) = (x,y^3)$ in $J^3(2,2)$. The 3-jet Z is v-sufficient in C^4 since for any C^4-realization f of Z, the varieties $Z^{-1}(0,0)$ and $f^{-1}(0,0)$ are locally homeomorphic. In fact, they both consist of the singleton point $(0,0)$. However Z is not C^0-sufficient in C^4 because the mapping g given by

$$g(x,y) = (x, y^3 + x^3 y)$$

is a C^4-realization of Z but not C^0-equivalent to Z. This can be seen by writing $Z = (Z_1, Z_2)$ and $g = (U,V)$ and then comparing the graphs of $Z_2(x,y) = y^3$ and $V(x,y) = y^3 + x^3 y$ (Figure 4.1) in $\mathbb{R}^3$. The same argument occurs in Example 1.2 of Chapter 1, shows that Z_2, and hence Z is not C^0-sufficient in C^4.

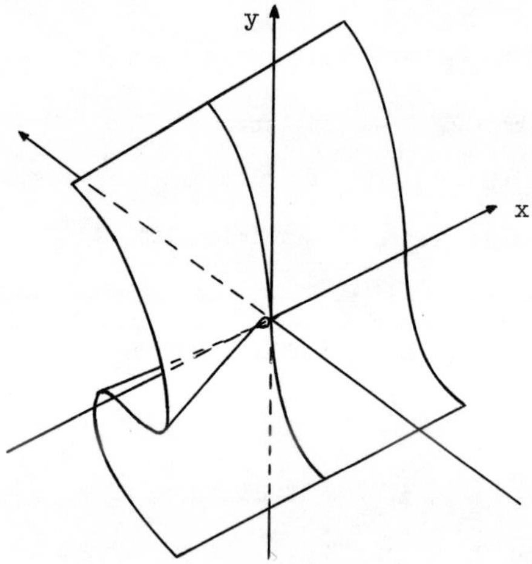

Figure 4.1. $V(x,y) = y^3 + x^3 y$

The study of $J^r(n,p)$, $p > 1$, with respect to C^0- and v-sufficiencies, is still in its initial stage. So far no criterion for C^0-sufficiency has been obtained. In the following we shall only mention some results about v-sufficiency of jets in this space.

Given vectors $v_1, \ldots, v_p$ in R^n, let h_i denote the distance from (the endpoint of) v_i to the subspace spanned by the vectors v_j, where $j \neq i$. If $p = 1$, write $h_1 = |v_1|$. Let $d(v_1, \ldots, v_p) = \min\{h_1, \ldots, h_p\}$, and note that $d(v_1, \ldots, v_p) = 0$ if and only if $v_1, \ldots, v_p$ are linearly dependent.

Given a mapping $f: \mathbb{R}^n \to \mathbb{R}^p$, $f(\bar{0}) = \bar{0}$, we define the <u>horn - neighborhood</u>

[29] of the variety $f^{-1}(0)$ of degree $d > 0$ and width $w > 0$ by

$$H_d(f;w) = \{x \in R^n \mid |f(x)| \leq w|x|^d\}.$$

Near the origin, this set is a horn-shaped set with vertex 0 and contains $f^{-1}(0) - \{0\}$ in its interior. For example, if $f(x,y) = x - y^2$, then $H_4(f;\frac{1}{2})$ is the

$$H_4(f;\tfrac{1}{2}) = \{(x,y) \in \mathbb{R}^2 \mid -\tfrac{1}{2}(x^2 + y^2)^2 \leq x - y^2 \leq \tfrac{1}{2}(x^2 + y^2)^2\}.$$

The following four theorems are proved by T. C. Kuo [30] in 1971.

<u>Theorem 4.1</u>. Let $Z = (Z_1,\dots,Z_p) \in J^r(n,p)$. Then the following conditions are equivalent:

(a) Z is v-sufficient in C^{r+1}.

(b) Given a polynomial map $g = (g_1,\dots,g_p)$ of degree $r+1$ with $j^{(r)}(g) = Z$, there exist $w > 0$, $\epsilon > 0$, $\delta > 0$, and a neighborhood U of $\overline{0} \in R^n$ such that

$$d(\text{grad } Z_1(x), \dots, \text{grad } Z_p(x)) \geq \epsilon|x|^{r-\delta}$$

for all x in $U \cap H_{r+1}(g;w)$.

(c) For any C^{r+1}-realization $f = (f_1,\dots,f_p)$ of Z, the variety $f^{-1}(\overline{0})$ admits $\overline{0}$ as a topologically isolated singularity, i.e. near $\overline{0} \in \mathbb{R}^n$, grad $f_1(x)$, ..., grad $f_p(x)$ are linearly independent everywhere on $f^{-1}(\overline{0}) - \{\overline{0}\}$.

<u>Theorem 4.2</u>. A jet $Z \in J^r(n,p)$ is not v-sufficient in C^{r+1} if and only if there exist an analytic arc β: $x_i = x_i(t)$, $x_i(0) = 0$, $i = 1,\dots,n$, and a C^{r+1}-realization f of Z such that the variety $f^{-1}(\overline{0})$ is everywhere singular along β.

In $J^r(n,1)$, since C^0-sufficiency and v-sufficiency are equivalent, the above theorem becomes a non-C^0-sufficiency criterion for an r-jet. As a special case, for $J^r(2,1)$, one has the following corollary (a short proof for this special case is in Kuo [29]).

Corollary 4.3. Let $f(x,y)$ be a given polynomial. Then there exists no r for which $j^{(r)}(f)$ is C^0-sufficient in C^{r+1} if and only if $f(x,y)$ is divisible by $(h(x,y))^2$, where $h(x,y)$ is a polynomial having zeros arbitrarily close to $\bar{0} \in \mathbb{R}^2$.

The following theorem was first announced by Thom ([73], Theorem 3).

Theorem 4.4. If a jet $Z \in J^{r-1}(n,p)$ is given, then for "almost all" p-tuples H_r of homogeneous polynomials of degree r, the r-jet $Z + H_r$ is visufficient in C^{r+1}.

Again the set of all such p-tuples H_r constitutes a Euclidean space and "almost all" means that the exception occurs on a proper algebraic subvariety of this space.

Theorem 4.5. If $f: \mathbb{R}^n \to \mathbb{R}^p$ is a local analytic mapping such that the variety $f^{-1}(\bar{0})$ has $\bar{0}$ as a topologically isolated singularity, then for all large r, $j^{(r)}(f)$ is v-sufficient in C^{r+1}.

If in $J^r(n,p)$ one considers v-sufficiency of r-jets in C^r, then all the above theorems will hold with δ replaced by 1 and $r + 1$ by r in the corresponding statements.

In the following we give a sufficient condition for an r-jet in $J^r(n,p)$ to be v-sufficient in C^ω, using the ratio test.

Theorem 4.6. Let $Z \in J^r(n,p)$ and suppose that $P: R^n \to R^p$ is any analytic mapping such that $j^{(r)}(P) = 0$. Then if $(Z + tP)^{-1}(\bar{0})$ satisfies the ratio test over the t-axis at $\bar{0}$, the r-jet Z is v-sufficient in C^ω.

Proof: By Theorem 5.1 in Chapter 5, we know that $(Z + tP)^{-1}(\bar{0})$ is regular over the t-axis at $\bar{0}$. Hence the mapping

$$Z + tP: (Z + tP)^{-1}(\bar{0}) \to R$$

is, by definition (see [47]), a controlled submersion. It is also clear that this mapping is proper. Then according to a theorem of Thom ([71], Theorem 1.G.1, p. 258), $(Z + tP)^{-1}(\overline{0})$ is locally trivial along the t-axis. In particular $Z^{-1}(\overline{0})$ and $(Z + P)^{-1}(\overline{0})$ are locally homeomorphic.

APPENDIX I

THOM'S THREE BASIC PRINCIPLES

In the introductory section of Chapter 4 we asserted that Thom's classification theorem for (stable) universal unfoldings is the key result in catastrophe theory. In this section we give a brief exposé of Thom's three basic principles in morphogenesis and indicate why the classification theorem is so fundamental. The first principle in morphogenesis asserts that the stability of any morphogenetic phenomenon, whether represented by a gradient system or not, is determined by the attractor set of a certain vector field. For a parametrized gradient vector field of the form $X_u = (V^u_{x_1}, \ldots, V^u_{x_n})$, where $V^u\colon \mathbb{R}^n \to \mathbb{R}$ is a smooth map depending smoothly on $u \in \mathbb{R}^r$, the attractors of X_u are precisely the sinks, i.e. the stable minima of V^u. Hence they lie in $M_V \subset \mathbb{R}^{n+r}$, which is defined to be given by $\text{grad}_x V = 0$. To be more precise in the formulation of this principle, we should define the following four terms -- a system, a state of a given system, a process of a given system, and a parameter.

(1) A system, for our purposes, is a collection or set, Q, of interesting qualities. (An interesting quality, however, is a primitive notion and is not defined.)

(2) A map from this set of qualities to $\mathbb{R}$ (or in a more general setting we could consider a map from Q to a finite dimensional Euclidean space), which assigns to each quality a unique real number, is a state of the system. The state space then is the set of all such maps and may thus be identified with the function space $\mathbb{R}^Q$, and hence with the Euclidean space $\mathbb{R}^{|Q|}$, where $|Q|$ is cardinality of the set Q. We usually identify the state with its image in $\mathbb{R}^Q$.

(3) A process of the system is a set of possible states and a rule (usually but not always given by a vector field) for selecting an actual state from possible states.

(4) A <u>parameter</u> (i.e. an r-dimensional parameter) for a process is an r-tuple of real numbers, on which the selection rule for determining the actual state of the process depends. (But the states of the system do not depend on the parameter.)

Thom's first principle can be phrased as soon as we know what a morphogenetic phenomenon is. This consists of a configuration space M (which will be an open subset of R^n for the purpose of this section, although it could be taken to be an n-dimensional compact manifold without boundary); an r-dimensional manifold U as parameter space; and a smooth vector field $X: U \times M \to TM$ where TM is the tangent bundle of M and $X_u: \{u\} \times M \to TM$ is the section map of X at u. Since the states have not yet been specified, this is not yet a process as defined in (3).

Then Thom asserts that, for each u, the subset of M corresponding to the attractor set (sometimes called the set of degenerate critical points or the set of unstable sources of X_u) is the set of possible states for the morphogenetic phenomonon at u. Nonetheless, those in which the system will lie are, of course, the set of attractors. The union over all u in U of these subsets, identified as a subset of $U \times M$, is the state space of the phenomenon. Then, if X is a parametrized gradient vector field, the state space is almost (but not quite) the set of M_V defined above when X is defined by $X(u,m) = -\nabla V^u(m) = (-\frac{\partial V^u}{\partial x_1}(m), \ldots, -\frac{\partial V^u}{\partial x_n}(m))$. Here $V: R^{n+r} \to R$ is smooth, $U = R^r$, $M = R^n$ (or open subsets of R^r, R^n respectively). Note that M_V also contains local maxima for V^u for various u and these are irrelevant for the application of the theory but not for the formalism (i.e., including them in M_V insures that M_V is a manifold). Hence for <u>generically</u> many such phenomena or processes (when the vector field is identified with its corresponding potential function, and when one is only considering processes determined by a vector field X of the form $-\nabla V$) the state space is an r-dimensional manifold embedded naturally in R^{n+r}. We will be concerned with the case $r = 4$ (although the case $r = 5$ works just as well).

By itself this is a rather weak statement because it is global in nature. When we combine it with the second defining property of $\mathfrak{F}_0$, we can give more precise information about the local structure of $V \in \mathfrak{F}_0$ near each such point in M_V.

Let $\overline{x}_0 = (\overline{p}_0, \overline{u}_0) \in \mathbb{R}^{n+4}$ be a fixed point in M_V where $\overline{p}_0 \in \mathbb{R}^n$, $\overline{u}_0 \in \mathbb{R}^4$, V our arbitrary generic "process." Because M_V is an 4-dimensional manifold, we can choose coordinates around $\overline{x}_0$ so that $\overline{x}_0 = (\overline{0}, \overline{0})$. Let U be a fixed neighborhood around $\overline{x}_0$ in R^{n+4}, on which $V|_U$ is a universal unfolding of the map $\tilde{f} = V|_{U \cap (\mathbb{R}^n \times \{\overline{u}_0\})}$. Now we invoke the essence of the classification of universal unfoldings. (We reemphasize here that universality of an unfolding is a local notion!) This says that $\tilde{f}$ can be reduced to one of seven polynomials*, the Thom polynomials. They are x^3, x^4, x^5, x^6, $x^3 + y^3$, $x^2 y + y^4$, $x^3 - 3xy^2$. To say that $\tilde{f}$ can be so reduced means that $\tilde{f}$ is locally <u>equivalent</u> to the sum of a nondegenerate quadratic form and exactly one of these seven polynomials, say g_i, $1 \leq i \leq 7$. Precisely, there is a local diffeomorphism φ fixing $\overline{x}_0$ such that $\tilde{f} \circ \varphi$ equals the sum of the two functions described, in some neighborhood of $\overline{x}_0$ in R^{n+r}. Moreover, since this is the case we know even more, we know that V is itself locally equivalent to the sum of the unique universal unfolding (of minimal codimension) of g_i, say $\tilde{g}_i$, and a quadratic form Q which may be assumed to be positive definite in an appropriate coordinate system.

In detail, one has the following. Let $V\colon S_0 \times W_0 \to \mathbb{R}$ where W_0 is an $\mathbb{R}^4$ neighborhood of $\overline{u}_0$ and S_0 is an $\mathbb{R}^n$ neighborhood of $\overline{p}_0$; then V can be reduced to one of the $\tilde{g}_i$ in $S_0 \times W_0$. Hence, V is equivalent to $h_i(\overline{x}, \overline{z}, \overline{u}, \overline{t}) = \tilde{g}_i(x,u,t) + \sum_1^q z_1^2$ where $\overline{u} = (u,v,w)$, $\overline{x}$ is x_1 of (x_1, x_2),

---

*If we consider oriented reduction [82], then $\tilde{f}$ can be oriented reduced to one of ten polynomials. They are x^3, $\pm x^4$, x^5, $\pm x^6$, $x^3 + y^3$, $x^2 y \pm y^4$, $x^3 - 3xy^2$. For $r = 5$ one includes the polynomials x^7, $x^2 y \pm y^5$, $x^3 + y^4$, for the case of oriented reduction one includes the polynomials x^7, $x^2 y \pm y^5$, $x^3 \pm y^4$.

$\overline{z} = (z_1,\ldots,z_q)$, depending on the corank, and $q = n - j$, $j = 1$ or 2 in $S_0 \times W_0$ and $\overline{g}_i$ is a constant unfolding of $\widetilde{g}_i$ of codimension 4. If we let j stand for either 1 or 2 in the following statements, it is sometimes quite confusing. Hence, from now on, we simply let $j = 2$. In reality, the letter j could be one sometimes. We are doing this artificial assignment of j simply because we feel that the reader can follow the arguments easier by fixing j instead of having j everywhere representing either one or two.

Since V is right equivalent to h_i for some i, $1 \leq i \leq 7$, we can find $\alpha\colon S_0 \times W_0 \to S_0$, $\psi\colon W_0 \to W_0$ and $\gamma\colon W_0 \to \mathbb{R}$ so that $V(\overline{x},\overline{z},\overline{u},t) = h_i(\alpha(\overline{x},\overline{z},\overline{u},t), \psi(\overline{u},\overline{t})) + \gamma(\overline{u},t)$ and $\alpha|_{S_0 \times \{\overline{u}_0\}} \equiv \alpha_{\overline{u}_0}\colon S_0 \to \mathbb{R}^n$ is nonsingular at $\overline{p}_0$ and ψ is a diffeomorphism in W_0. From this we have that, if

$$\alpha(\overline{x},\overline{z},\overline{u},t) = (\alpha_1(\overline{x},\overline{z},\overline{u},t), \alpha_2(\overline{x},\overline{z},\overline{u},t), \alpha_3(\overline{x},\overline{z},\overline{u},t), \ldots, \alpha_n(\overline{x},\overline{z},\overline{u},t))$$

then

$$\begin{pmatrix} \frac{\partial V}{\partial x_1}(\overline{x},\overline{z},\overline{u},t) \\ \frac{\partial V}{\partial x_2}(\overline{x},\overline{z},\overline{u},t) \\ \frac{\partial V}{\partial z_1}(\overline{x},\overline{z},\overline{u},t) \\ \cdot \\ \cdot \\ \cdot \\ \cdot \\ \frac{\partial V}{\partial z_q}(\overline{x},\overline{z},\overline{u},t) \end{pmatrix} = \Big(\frac{\partial \widetilde{g}_i}{\partial x_1}(\alpha(\overline{x},\overline{z},\overline{u},t), \psi(\overline{u},y)), \frac{\partial \widetilde{g}_i}{\partial x_2}(\alpha(\overline{x},\overline{z},\overline{u},t), \psi(\overline{u},t)), \; 2\alpha_3(\overline{x},\overline{z},\overline{u},t),\ldots,2\alpha_n(\overline{x},\overline{z},\overline{u},t)\Big)\cdot J$$

where

$$J = \begin{pmatrix} \frac{\partial \alpha_1}{\partial x_1}, & \frac{\partial \alpha_1}{\partial x_2}, & \frac{\partial \alpha_1}{\partial z_1}, & \ldots, & \frac{\partial \alpha_1}{\partial z_q} \\ \cdot & \cdot & \cdot & \cdot & \cdot \\ \frac{\partial \alpha_n}{\partial x_1}, & \frac{\partial \alpha_n}{\partial x_2}, & \frac{\partial \alpha_n}{\partial z_1}, & \ldots, & \frac{\partial \alpha_n}{\partial z_q} \end{pmatrix} (\overline{x},\overline{z},\overline{u},t) .$$

For $(\overline{u},t)$ near $\overline{u}_0 = (\overline{0},\overline{0})$ in W_0, $\alpha_{(\overline{u},t)}: S_0 \to \mathbb{R}^n$ sending $(\overline{x},\overline{z}) \to \alpha(\overline{x},\overline{z},\overline{u},t)$ is a diffeomorphism.

Thus the vector $\left(\frac{\partial V}{\partial x_1}, \frac{\partial V}{\partial x_2}, \frac{\partial V}{\partial z_1}, \ldots, \frac{\partial V}{\partial z_q}\right)_{(\overline{x},\overline{z},\overline{u},t)} = (\overline{0})$ if and only if the vector

$$\left(\frac{\partial \widetilde{g}_i}{\partial x_1}(\alpha(\overline{x},\overline{z},\overline{u},t) , \psi(\overline{u},t)) , \frac{\partial \widetilde{g}_i}{\partial x_2}(\alpha(\overline{x},\overline{z},\overline{u},t) , \psi(\overline{u},t)) , 2\alpha_3(\overline{x},\overline{z},\overline{u},t) , \ldots, 2\alpha_n(\overline{x},\overline{z},\overline{u},t)\right) = 0 ,$$

since the matrix $J = D_{(\overline{x},\overline{z})}\alpha(\overline{x},\overline{z},\overline{u},t)$, the differential of α at $(\overline{x},\overline{z})$, is nonsingular. And this says that if $(\overline{u},t)$ is fixed, $(\overline{x},\overline{z})$ is a singular point for $V^{(\overline{u},t)}$ if and only if

$$\alpha(\overline{x},\overline{z},\overline{u},t) = (\alpha_1(\overline{x},\overline{z},\overline{u},t) , \alpha_2(\overline{x},\overline{z},\overline{u},t) , 0 , \ldots, 0)$$

is a singular point for $h_i^{\psi(\overline{u},t)}$, which is a mapping $S_0 \to \mathbb{R}$, parametrized by $\psi(\overline{u},t)$.

Thus, for any point to be a singular point of h_i at $\psi(\overline{u},t)$ it must have zero components in the variable $\overline{z} = (z_1,\ldots,z_q)$.

Now, let us consider the relationship of h_i to $\overline{g}_i$:

$$h_i(\overline{x},\overline{z},\overline{u},t) = \overline{g}_i(\overline{x},\overline{u},t) + Q(\overline{z}) .$$

So,

$$\frac{\partial h_i}{\partial x_1}(\bar{x},\bar{z},\bar{u},t) = \frac{\partial \bar{g}_i}{\partial x_1}(\bar{x},\bar{u},t)$$

$$\frac{\partial h_i}{\partial x_2}(\bar{x},\bar{z},\bar{u},t) = \frac{\partial \bar{g}_i}{\partial x_2}(\bar{x},\bar{u},t) .$$

Thus, $(\alpha_1(\bar{x},\bar{z},\bar{u},t), \alpha_2(\bar{x},\bar{z},\bar{u},t), \bar{0})$ is a singular point of h_i at $\psi(\bar{u},t)$ if and only if $(\alpha_1(\bar{x},\bar{z},\bar{u},t), \alpha_2(\bar{x},\bar{z},\bar{u},t))$ is a singular poing of $\bar{g}_i$ at $\psi(\bar{u},t)$.

Thus, we have this conclusion expressed succinctly as the following: Let

$$\bar{B} = \{(\bar{x},\bar{u},t) \in \pi(S_0) \times W_0 : \frac{\partial \bar{g}_i}{\partial x_1}(\bar{x},\bar{u},t) = \frac{\partial \bar{g}_i}{\partial x_2}(\bar{x},\bar{u},t) = 0\}$$

and

$$\bar{A} = \{(\bar{x},\bar{z},\bar{u},t) \in S_0 \times W_0 : \frac{\partial V}{\partial x_1} = \frac{\partial V}{\partial x_2} = \frac{\partial V}{\partial z_1} = \cdots = \frac{\partial V}{\partial z_q} = 0\}$$

Here, $\pi : S_0 \to \mathbb{R}^2$ (in fact $\mathbb{R}^j$) is the projection onto the first 2 (j respectively) coordinates. Then if

$$\Phi(\bar{x},\bar{z},\bar{u},t) = (\alpha_1(\bar{x},\bar{z},\bar{u},t), \alpha_2(\bar{x},\bar{z},\bar{u},t), \psi(\bar{u},t)) ,$$

we have

$$\Phi^{-1}(\bar{B}) = \bar{A}$$

in $S_0 \times W_0$.

From the identity concerning the first order partials of V and h_i, we also derive an identity for the second order partials of V on $\bar{A}$ and g_i on $\bar{B}$.

Then we have the identity between $n \times n$ matrices:

$$\begin{pmatrix} \frac{\partial^2 V}{\partial x_1^2} & \frac{\partial^2 V}{\partial x_1 \partial x_2} & \frac{\partial^2 V}{\partial x_1 \partial z_1} & \cdots & \frac{\partial^2 V}{\partial x_1 \partial z_q} \\ \cdots & \cdots & \cdots & \cdots & \cdots \\ \cdots & \cdots & \cdots & \cdots & \cdots \\ \frac{\partial^2 V}{\partial z_q \partial x_1} & & & & \frac{\partial^2 V}{\partial z_q^2} \end{pmatrix} (\bar{x},\bar{z},\bar{u},t)$$

$$= J^t(\bar{x},\bar{z},\bar{u},t) \begin{pmatrix} \frac{\partial^2 h_i}{\partial x_1^2} & \cdots & \frac{\partial^2 h_i}{\partial x_1 \partial z_q} \\ \cdots & \cdots & \cdots \\ \cdots & \cdots & \cdots \\ \frac{\partial^2 h_i}{\partial z_q \partial x_1} & \cdots & \frac{\partial^2 h_i}{\partial z_q^2} \end{pmatrix} \Phi(\bar{x},\bar{z},\bar{u},t)$$

where J is the Jacobian matrix of α with respect to the $(\bar{x},\bar{z})$ coordinates (as on the previous page).

By definition of h_i, however, the Hessian of h_i at a point $(\bar{x},\bar{z},\bar{u},t)$ in $\bar{A}$ look like

$$\left(\begin{array}{cc|cccc} \frac{\partial^2 h_i}{\partial x_1^2}, & \frac{\partial^2 h_i}{\partial x_1 \partial x_2} & & & 0 & \\ \frac{\partial^2 h_i}{\partial x_2 \partial x_1}, & \frac{\partial^2 h_i}{\partial x_2^2} & & & & \\ \hline & & 2 & & & \\ & & & 2 & & \\ & 0 & & & \ddots & \\ & & & & & 2 \end{array}\right) (\alpha_1(\bar{x},\bar{z},\bar{u},t),\alpha_2(\bar{x},\bar{z},\bar{u},t),\bar{0},\psi(\bar{u},t))$$

and this matrix clearly is the same matrix as

$$\begin{pmatrix} \begin{matrix} \frac{\partial^2 \bar{g}_i}{\partial x_1^2}, & \frac{\partial^2 \bar{g}_i}{\partial x_1 \partial x_j} \\ \frac{\partial^2 \bar{g}_i}{\partial x_j \partial x_1} & \frac{\partial^2 \bar{g}_i}{\partial x_j^2} \end{matrix} & 0 \\ 0 & \begin{matrix} 2 & & \\ & \ddots & \\ & & 2 \end{matrix} \end{pmatrix}_{(\alpha_1(\bar{x},\bar{z},\bar{u},t),\alpha_2(\bar{x},\bar{z},\bar{u},t)\bar{0},\psi(\bar{u},t))}$$

From this identity, we wish to derive two conclusions:

(1) Universality of the catastrophe set: Let $K_V = \{(u,v,w,t) \in W_0 \colon V^{(u,v,w,t)}$ has a degenerate critical point in $S_0\}$ and $K_{\bar{g}_i} = \{(u,v,w,t) \colon \bar{g}_i(u,v,w,t)$ has a degenerate critical point in $\pi(S_0)\}$. Then $\psi^{-1}(K_{\bar{g}_i} \cap W_0) = K_V$.

Since $\tilde{g}_i$ can be considered as $\bar{g}_i(x_1,x_2,\tilde{\psi}(u,v,w,t))$ where $\tilde{\psi}(u,v,w,t) = (u_1,\ldots,u_k)$, $k \leq 4$, given the unfolding parameters for $\tilde{g}_i$, then $K_{\bar{g}_i} = \tilde{\psi}^{-1}(K_{\tilde{g}_i})$ so that we have $K_V = (\tilde{\psi} \circ \psi)^{-1}(K_{\tilde{g}_i} \cap W_0)$. This relation enables us to relate, that is, keep track of, the positions in space-time near the origin $(\bar{u}_0)$ of the process, i.e., in a neighborhood W_0 of $\bar{u}_0$, at which there is a degenerate critical point in S_0 which "determines" the state of the process V. K_V is called the <u>catastrophe set</u> of the process for V and $K_{\tilde{g}_i}$ is called a <u>universal catastrophe set</u>.

As we shall see, shortly, what we "observe" of a process is in fact K_V, according to Thom's formulation of a morphogenetic phenomenon.

(2) Universality of nondegenerate minima: We know that V is right equivalent to h_i. Thus, a point p with coordinates $(\bar{x},\bar{z})$ is a local non-degenerate (i.e. stable) minimum of $V^{(\bar{u},t)}$ if and only if $\alpha(\bar{x},\bar{z},\bar{u},t)$ is a local nondegenerate minimum of $h_i^{\psi(\bar{u},t)}$. The latter statement is equivalent to:

$$\pi \circ \Phi(p) = \pi \circ \Phi(\bar{x},\bar{z},\bar{u},t) = (\alpha_1(\bar{x},\bar{z},\bar{u},t),\alpha_2(\bar{x},\bar{z},\bar{u},t))$$

is a local nondegenerate minimum of $\bar{g}_i^{\psi(\bar{u},t)}$. Now, $\pi \circ \Phi(\bar{x},\bar{z},\bar{u},t)$ is a local nondegenerate minimum of $\bar{g}_i^{\psi(\bar{u},t)}$ if and only if $(\alpha_1(\bar{x},\bar{z},\bar{u},t),\alpha_2(\bar{x},\bar{z},\bar{u},t))$ is a local nondegenerate minimum of $\tilde{g}_i^{(\tilde{\Psi}\circ\psi)(\bar{u},t)}$.

All of these statements hold because for $(\bar{u},t)$ near $(0,0,0,0)$ (i.e. $\bar{u}_0$) in W_0, $\alpha|_{S_0 \times \{(\bar{u},t)\}}$ is a diffeomorphism onto S_0.

Thus, we are able, for each $(\bar{u},t) \in W_0$, to keep track of degenerate and nondegenerate minima of V-points with coordinates $(\bar{x},\bar{z})$ in S_0 by knowing α_1,α_2 since $(\alpha_1(\bar{x},\bar{z},\bar{u},t),\alpha_2(\bar{x},\bar{z},\bar{u},t))$ are coordinates in $\pi \circ \alpha(S_0)$ for a minimum of the same type of degeneracy or nondegeneracy as that of h_i at $\tilde{\Psi}\circ\psi(\bar{u},t)$. Conversely, if (x_1,x_2) are coordinates at $(u_1,\ldots,u_k) \in \mathbb{R}^k$ of a minimum of $\tilde{g}_i(x_1,x_2,u_1,\ldots,u_k)$, then $(x_1,x_2,\bar{0})$ are coordinates of a minimum (stable or unstable) for h_i at any parameter point in $(\tilde{\Psi}\circ\psi)^{-1}(u_1,\ldots,u_k)$ in $\mathbb{R}^4$.

Now, $\alpha|_{S_0 \times \{(\bar{u},t)\}}$ is one-to-one. So there is a unique point $p(\bar{u},t) = (\bar{x},\bar{z},\bar{u},t)$ so that $\alpha(p(\bar{u},t)) = (x_1,x_2,\bar{0})$. Here $(\bar{u},t) \in (\tilde{\Psi}\circ\psi)^{-1}(u_1,\ldots,u_k)$. Then, $p(\bar{u},t)$ is a minimum (stable or unstable) of $V^{(\bar{u},t)}$ in S_0. Thus, we do have (local) universality of the local minima of any process V which reduces to $\tilde{g}_i$ in that knowing the configuration in R^2 (or R^1) of stable (unstable) local minima for $\tilde{g}_i$ enables one to find (with the aid of α and $\tilde{\Psi}\circ\psi$) the local configuration in M_V near $\bar{x}_0 = (\bar{p}_0,\bar{u}_0)$ of the stable (unstable) local minima for $V^{(\bar{u},t)}$ for $(\bar{u},t) \in W_0$.

Hence, we can conclude that $W_0 \cap M_V$, the set of equilibrium points containing the sets of states determined by V, is diffeomorphic to a suspension of $W_0 \cap M_{\tilde{g}_i + Q}$. (The suspension is necessary as we have seen because of a possibly larger number of unfolding parameters in V than in $\tilde{g}_i$. Thus, the local study of M_V reduces to the local study of $M_{\tilde{g}_i + Q}$. And it is easy to see that the local study of this set is equivalent to the local study of $M_{\tilde{g}_i}$, which is a manifold of dimension equal to the codimension of the Thom polynomial which it unfolds.

This reduction lies at the heart of Thom's theory of morphogenesis. We formulate this reduction as his second principle: What is interesting about morphogenesis, locally, is the transition, as the parameter varies, from a stable state of the vector field X to an unstable state and back to a stable state by means of the process which we use to model the system's local morphogenesis.

In the context of a gradient vector field $X = -\nabla V$, we can say that the variation (as u varies) between nondegenerate (stable) and degenerate (unstable) minima of the corresponding parametrized potential function determines the local morphogenesis of the gradient system. Thus, we envision some type of parametrized flow on the manifold M_V which picks out states and which characterizes the local morphogenesis of the system described by V by means of some of its discontinuities.

But M_V is globally too arbitrary. What the reduction mentioned above implies is that, locally, we may study M_V by considering $M_{\tilde{g}_i + Q}$ (and thus $M_{\tilde{g}_i}$), and do not lose any important morphogenetic information. Thus, we are really concerned with local "minima selection" on only 7 manifolds in order to characterize the local morphologies of a generic set of gradient systems*.

The amount of qualitative, as opposed to quantitative, description that enters now in our local study depends heavily upon the process to be studied. Moreover, the unfolding parameters themselves assume a physical meaning of importance as "control parameters." Indeed, a control parameter of a process shall be defined as a parameter appearing in the universal unfolding (of minimal dimension) to which V reduces.

The procedure, then, may be outlined as follows. First, we determine whether a particular system is describable as a gradient system and if possible we correlate the potential function with a physically relevant (and, in the best cases mathematically expressible) function. If one is dealing with a gradient system (of sufficient smoothness), the next question is to identify (by means of an understanding or knowledge of the system under study) the relevant control

---

*Cheerful remark: Any time we go from infinity to 7, we have achieved a great deal.

parameters (e.g. space-time coordinates in the case of biology). If the number r of such control parameters satisfies $r \leq 4$ (we might, but did not, allow $r \leq 5$), then one can use genericity (or stability) arguments to assert that V will reduce to one of the seven unfoldings described above. The ones to be tried, of course, should be those among the seven with the same number r of unfolding parameters as the number of control parameters. If this is not known, this step is, of course, impossible. But in the example described in section 4 of Chapter 4, it is. One can then model the state changes as a path on a selected subset of the seven catastrophe sets M_{g_i}. The one most commonly used so far as a first approximation, because it appears in the higher codimensional cases (and is simplest to calculate), is the cusp catastrophe corresponding to the polynomials $\frac{1}{4}x^4$. Here

$$\tilde{g}(x,u,v) = \frac{1}{4}x^4 + \frac{1}{2}ux^2 + vx$$

is the universal unfolding of $\frac{1}{4}x^4$ and $M_{\tilde{g}} = \{(x,u,v): -v = x^3 + ux\}$. Thus on this 2-dimensional manifold we envision the possible paths which the system must follow.

To state the third principle, we need to recall for $V: R^{n+4} \to R$, interpreted as a global process, the set

$$K_V = \{(\bar{x},\bar{u}): \frac{\partial V}{\partial x_1}(\bar{x},\bar{u}) = \frac{\partial V}{\partial x_2}(\bar{x},\bar{u}) = 0 = \det(\frac{\partial^2 V}{\partial x_1 \partial x_j})(\bar{x},\bar{u})\}$$

which consists of the parametrized degenerate critical points of the system described by $-\nabla V = X$.

By virtue of the nature of the reduction process, locally (i.e. in an appropriate neighborhood of each point in K_V and the corresponding image point in dom $\tilde{g}_i$) K_V had the same geometric structure as $K_{\tilde{g}_i}$ (defined similarly as K_V) since V reduces to $\tilde{g}_i$. Moreover, if $\pi: K_V \to \mathbb{R}^4$ is the natural projection and $\mathcal{K}_V = \pi(K_V)$, then $\mathcal{K}_V$ has the same local geometrical structure as a "stratified" set that $\mathcal{K}_{\tilde{g}_i} = \pi(K_{\tilde{g}_i}) \subset \mathbb{R}^k$ (k = unfolding codimension of $\tilde{g}_i$)

has for a generic set of V.

Rigorously, one has then the following. Our sequence of observations above yields the conclusions that if our process V reduces to $\tilde{g}_i$, then there will be a map $\hat{\psi} = \tilde{\psi} \circ \psi: W_0 \to \mathbb{R}^k$ so that $\mathcal{K}_V \cap W_0 = \hat{\psi}^{-1}(\mathcal{K}_{\tilde{g}_i}) \cap W_0$.

If we could then decompose $\mathcal{K}_{\tilde{g}_i}$ into a union of r disjoint manifolds $\bigcup_1^r A_\ell$, then by the transversality theorem, $\mathcal{K}_V$ would be the disjoint union of r disjoint manifolds $\cup \hat{\psi}^{-1}(A_\ell)$ if $\hat{\psi}$ is transverse to each of the r strata $A_1, \ldots, A_r$.

Moreover, if codimension $A_\ell = n_\ell$ in $\mathbb{R}^4$, then in $\mathbb{R}^4$ $\operatorname{cod} \hat{\psi}^{-1}(A_\ell) = n_\ell$ also. Hence, we know that $\mathcal{K}_V \cap W_0$ can be "stratified" or decomposed into a disjoint union of r manifolds each of the same relative size in $\mathbb{R}^4$ as their images are in $\mathbb{R}^4$ under $\hat{\psi}$. So, we know that as a subset of space time near an arbitrarily defined origin, the set of positions $(\bar{u}, t)$ at which the state of the system is at a degenerate, i.e., unstable critical position, is formed by a union of r such manifolds as indicated.

According to Thom, for any* morphogenetic process unfolding, that is, occuring in space time, and which is describable by a global potential function V, the local morphology of the process, that is, the evolution of the geometrical and/or topological form or properties of the substrata in which the process is occurring, is determined by the topology of $\mathcal{K}_V$ in W_0. To get a better feeling for what we really would see then over time, one should look at various time sections of $\hat{\psi}$. At each instant measured by t, we have an induced set $\hat{\psi}_t^{-1}(\mathcal{K}_{\tilde{g}_i}) = \mathcal{K}_V \cap W_0 \cap (\mathbb{R}^3 \times \{t\})$ where $\hat{\psi}_t: W_0 \cap (\mathbb{R}^3 \times \{t\}) \to \mathbb{R}^k$.

Generically, $\hat{\psi}$ is transverse to $\mathcal{K}_{\tilde{g}_i}$, i.e. $\hat{\psi}$ is transverse to each stratum of $\mathcal{K}_{\tilde{g}_i}$. As such, one can show that the set $\{t: \hat{\psi}_t$ is not transverse to $\mathcal{K}_{\tilde{g}_i}\}$ is a set of measure zero. That is, we can in theory a priori reconstruct visually $\mathcal{K}_V$ from its sections $\mathcal{K}_V^t$ just by knowing $\mathcal{K}_{\tilde{g}_i}$ and the transversality of $\hat{\psi}$ to the strata $\{A_1, A_2, \ldots, A_r\}$ of $\mathcal{K}_{\tilde{g}_i}$.

*I.e., for a generic set of processes; cf. below.

That $\hat{\psi}_t$ is transverse to $\mathcal{K}_{\tilde{g}_i} = \bigcup_1^r A_\ell$ now implies that if $\operatorname{cod}(A_\ell) = n_\ell$ in $\mathbb{R}^k$ then $\operatorname{cod}(\hat{\psi}_t^{-1}(A_\ell)) = n_\ell$ not in $\mathbb{R}^4$ but $\mathbb{R}^3$. So the size of the strata in $\mathbb{R}^3$ would be reduced by one. The transversality of $\hat{\psi}_t$ then suggests that the stratified structure of the shock wave at each instance of observation will be the same over various intervals of time. Moreover, this structure will mirror the universal stratified structure of $\mathcal{K}_{\tilde{g}_i}$. Thus, one could <u>in theory</u> piece together this time parametrized structure to determine which universal catastrophe set is being observed through the intermediary of $\hat{\psi}$ and its time level maps $\{\hat{\psi}_t\}$. But, of course, in practice this requires considerable ingenuity and geometric imagination for any of the sets $\mathcal{K}_{\tilde{g}_i}$ more complicated than the cusp.

There is a considerable (local) improvement upon this theory, developed by G. Wasserman and called (r,s) stability. In the $(3,1)$ case, one has the following: If V is $(3,1)$ right equivalent to one of the finitely many $(3,1)$ stable unfoldings g (at $(\overline{0},(\overline{0},0)) \in \mathbb{R}^{n+3+1}$), then there are germs of diffeomorphisms $\rho\colon (\mathbb{R},0) \to (\mathbb{R},0)$, and, for t near 0 in $\mathbb{R}$ and (u,v,w,t) near $\overline{0}$ in $\mathbb{R}^4$, $\varphi_t\colon (\mathbb{R}^3,0) \to (\mathbb{R}^3,0)$ and $\theta_{(u,v,w,t)}\colon (\mathbb{R}^n,0) \to (\mathbb{R}^n,0)$ such that

$$V(\overline{x},(u,v,w),t) = g(\theta_{(u,v,w,t)}(\overline{x}),\varphi_t(u,v,w),\rho(t)) .$$

Going through the derivation as above, the map $\hat{\psi}$ in this case has the form $\hat{\psi}(u,v,w,t) = (\varphi(u,v,w,t),\rho(t))\colon (\mathbb{R}^4,0) \to (\mathbb{R}^4,0)$.

Hence, $\hat{\psi}_{t_0}$ maps time levels to time-levels and preserves topological structure. Although there is no global classification of $(3,1)$ stable unfoldings, nor, is there even a finite classification of $(1,3)$ stable unfoldings, one clearly should incorporate $(3,1)$ theory [83] into an analysis of local catastrophic phenomena. After all, the coordinate changes are so nice. One does not 'flip-flop' space and time coordinates in modelling real phenomena; but in general, this could not be prevented in the original Thom-Zeeman-Mather-Mathematical framework presented here.

Thom gives a metascientific description of the morphological significance of the catastrophe set as follows: the stable minima occur at points $(\overline{u},t)$ in a fixed component of the complement of K_V in W_0. If, for example, we fix $\overline{u}$ to lie in a very small neighborhood of $\overline{0}$ in R^3 and allow t to increase, the corresponding stable minima which now correspond to states of the system at the points $(\overline{u},t)$, may approach a degenerate minimum p_0, at $(\overline{u},t_0)$ for $V^{(u,t_0)}$. As the states transverse p_0, the sudden shift from stable state to unstable state back to stable state produces a shock which we will see in that part of the substrata near $(\overline{u},t_0)$. What we mean by a "shock" is that there will be a visible division of the space-time values, i.e., regions of the substrata, near $(\overline{u},t_0)$ where the dividing boundaries will be given by space-time coordinates corresponding to degenerate minima and the zones divided correspond to nondegenerate stable minima. These latter zones of stability are labelled, by Thom, "chreods" [74]. Thus, Thom's third principle states that what is observed in a process undergoing morphogenesis is precisely the shock wave and resulting configuration of chreods separated by the strata of the shock wave, at each instant of time (in general) <u>and</u> over intervals of observation time.

The universality of both $M_{\tilde{g}_i}$ and K_{g_i}, described above, now yield the fact that the morphology described as occurring in space-time is mirrored "universally" and topologically in the space of unfolding parameters for the universal unfolding $\tilde{g}_i$ to which V reduces on the right. Thus, Thom draws the remarkable conclusion that we can study (topologically) the local morphology of "any" process V by studying the corresponding local morphology for $\tilde{g}_i$. This explains why the elementary catastrophes, that is, the sets $K_{\tilde{g}_i}$ as well as the polynomials g_i and $\tilde{g}_i$ have become of such interest.

What we observe locally, therefore, of the process described globally by V will be topologically the same as what we observe locally of the process described by $\tilde{g}_i$.

As promised, we should say a few words about the meaning of the term "observe." If the object is embedded in a medium (say an embryo), then we will observe "shock" waves at points in space-time at which the evolutionary state path meets the set $\mathcal{K}_V$.

On the other hand, if we have specified the unfolding parameters to be control parameters with some specific meaning or measurement attached to them, the model would predict evident discontinuities in the evolution of the system at those control parameters values lying in $\mathcal{K}_V$.

Thus, to classify an observed phenomenon or to support a hypothesis about the local underlying dynamic (i.e. $X = -\nabla V$), we need in principle only observe the process, study geometrically the observed "catastrophe (discontinuity) set" and try to relate it to one of the finitely many universal catastrophe sets, which would then become our main object of interest.

APPENDIX II

THE PROOF OF THOM'S CLASSIFICATION THEOREM

The proof of Thom's Classification Theorem is deeply involved with the Theorem of the Residual Singularity (Theorem 5.3 of Chapter 3, it is also often referred as the Splitting Lemma), which can be reinterpreted as follows: Let $f \in m(n)^2$ have corank p, then there is $g \in m(p)^3$ such that

$$f(x_1,\dots,x_n) \sim g(x_1,\dots,x_p) + q(x_{p+1},\dots,x_n)\ ,$$

where q is a non-degenerate quadratic form.

In fact, if

$$h(x_1,\dots,x_n) = g(x_1,\dots,x_p) + q(x_{p+1},\dots,x_n)\ ,$$

where $h \in m(n)^2$, $g \in m(n)^3$ and q is a non-degenerate quadratic form and if G is a universal unfolding of g, then $G + q$ is a universal unfolding of h. To see this we reduce q to a sum of squares, where it is obvious that $m(n)/\langle\frac{\partial h}{\partial x}\rangle$ and $m(p)/\langle\frac{\partial g}{\partial x_1},\dots,\frac{\partial g}{\partial x_r}\rangle$ have the same basis. Recall that if $\{b_1,\dots,b_r\}$ is a basis for $m(n)/\langle\frac{\partial h}{\partial x}\rangle$, where $h \in m(n)^2$, then $h(x) + \sum_{i=1}^{r} u_i b_i(x) = H(x,u)$ is a universal unfolding of h. Thus $G + q$ is a universal unfolding of h in case of G is a universal unfolding of g. Therefore the Splitting Lemma allows us to concentrate on degenerate germs in computing universal unfoldings.

From Lemma 3.1 of Chapter 4, we realize that the corank of f is important since we prove in this lemma that if $\operatorname{corank} f \geq 3$, then $\operatorname{codim} f \geq 6$. This result, therefore, allows us to concentrate upon germs f of corank 1 or 2 since we are only interested in those f with $\operatorname{codim} f \leq 4$.

(I) corank $f = 1$. We claim that f reduces (on the right) to x^i for $3 \leq i \leq 6$.

By the Splitting Lemma, we know that f reduces to a germ $h \in m(1)^3$ such that $\operatorname{codim} h \leq 4$. This implies that h is finitely determined, in other words, for some k, $\langle \frac{dh}{dx} \rangle_{\varepsilon(1)} \supset m(1)^k$ and thus h is not flat at 0. (A germ h is flat at 0 if $\frac{d^m h}{dx^n} = 0$ for all $m = 1, 2, \ldots$) Hence,

$$h(x) = \frac{d^{\ell} h}{dx^{\ell}}(0)x^{\ell} + x^{\ell+1}\theta(x),$$

where $\frac{\theta(x)}{|x|} \to 0$ as $x \to 0$ and $\ell \geq 3$. Therefore $\theta \in m(1)$. As a consequence, we can write h in the form,

$$h(x) = \pm x^{\ell}(a \pm \theta(x)).$$

We choose the sign of x^{ℓ} so that $a > \pm \theta(x)$ in a neighborhood of 0. Then let

$$\varphi(x) = x(\sqrt[\ell]{a \pm \theta(x)}).$$

Clearly $\varphi \in L(1)$ and $h(x) = \pm (\varphi(x))^{\ell}$. Hence g is (right) equivalent to $g(x) = \pm x^{\ell}$. Thus, $\operatorname{codim} h = \operatorname{codim} g \leq 4$. But

$$\dim_{\mathbb{R}} m(1)/\langle x^{\ell-1} \rangle = \dim_{\mathbb{R}} \langle x, x^2, \ldots, x^{\ell-2} \rangle = \ell - 2 \leq 4$$

implying that $3 \leq \ell \leq 6$.

If ℓ is odd, $\psi(x) = -x$ yields the conclusion that $g(x)$ is equivalent to $-g(x)$. So far, we have completed the discussion of the germs $g_1, g_2, g_3, g_4, g_8, g_9$ in the list in Theorem 5.5 of Chapter 3.

(II) The interesting (and longer) part of the proof is concerned with the case corank $f = 2$. In this case, f reduces to a germ $\tilde{h} \in m(2)^3$, by the Splitting Lemma. So we should investigate $\tilde{h}$ and we do this by considering the 3-jet of $\tilde{h}$, $j^{(3)}(\tilde{h})$, which is therefore a homogeneous polynomial in two variables, x and y, of degree 3. Let $P(x,y) = j^{(3)}(\tilde{h})$ be this cubic. The idea of the proof now is to investigate the types and number of orbits of cubic polynomials in two variables under a (right) equivalent group action, that is, action on the right by $L^{(3)}(2)$. Fortunately, there are only finitely many such orbits in this relatively low dimensional case. What we hope then is that in each such orbit there is a 3-jet which is 3-determined. We can then conclude that $\tilde{h}$ is equivalent to a realization h of this 3-determined jet and hence, f reduces on the right to h.

Therefore, let us look at a real cubic form

$$P(x,y) = \tilde{a}_1x^3 + \tilde{a}_2x^2y + \tilde{a}_3xy^2 + \tilde{a}_4y^3 .$$

This corresponds to a 3-jet of a germ $\tilde{h} \in m(2)^3$, that is, the Taylor expansion of degree three of $\tilde{h}$ at (0,0). We can dehomogenize P by letting $X = \frac{x}{y}$, $Y = y$. Then

$$P(X,1) = \tilde{a}_1X^3 + \tilde{a}_2X^2 + \tilde{a}_3X + \tilde{a}_4 = \tilde{a}_1(X - \alpha_1)(X - \alpha_2)(X - \alpha_3) ,$$

for $\alpha_1,\alpha_2,\alpha_3 \in \mathbb{C}$ and at least one α_i, say α_1, in $\mathbb{R}$. Thus $P(X,1)$ splits into three factors implying that $P(x,y)$ splits into 3 factors

$$P(x,y) = (a_1x + b_1y)(a_2x + b_2y)(a_3x + b_3y) .$$

Then, the type of roots that $P(X,1)$ has yields different factorizations for P and thus yields an interpretation of the linear independence or dependence over $\mathbb{C}$ of the three vectors (a_1,b_1), (a_2,b_2), (a_3,b_3) in $\mathbb{C}^2$. The cases are:

(a) $P(X,1)$ has three distinct roots. This is the same as saying that $P(x,y) = (a_1x + b_1y)(a_2x + b_2y)(a_3x + b_3y)$, and the three factors (a_1,b_1), (a_2,b_2), (a_3,b_3) are pairwise linearly independent over $\mathbb{C}$. We can further classify this case into two subcases:

(i) $P(X,1)$ has three distinct real roots.

(ii) $P(X,1)$ has one real and two complex and thus conjugate roots.

(b) P has a double root, i.e. $P(X,1) = (X - \alpha_1)(X - \alpha_2)^2$. This is the same as saying that

$$P(x,y) = (a_1x + b_1y)(a_2x + b_2y) .$$

Thus, two of the vectors are linearly independent over C and the other is a nonzero multiple of one of the two. Since P is a real polynomial, $\alpha_1 \in \mathbb{R}$ and so $\alpha_2 \in \mathbb{R}$. Hence (a_1,b_1) and (a_2,b_2) are <u>real</u> vectors.

(c) P has a triple real root, i.e.

$$P(x,y) = (a_1x + b_1y)^3 ,$$

thus for the three real vectors, no two of them are linearly independent and none is the zero vector.

To have a complete classification, we also should add to our three cases a fourth trivial case wherein

(d) $\tilde{h} \in m(2)^4$, in this case $P \equiv 0$.

<u>Schema for the Various Cases</u>

Case (aI): The polynomial $P(X,1)$ has three distinct real roots. In this case each $a_i,b_i \in \mathbb{R}$, $i = 1, 2, 3$. Consider an element $T \in G\ell(2,\mathbb{R})$ as a 3-jet. In particular, let

$$T = \begin{pmatrix} a_1 & b_1 \\ a_2 & b_2 \end{pmatrix} .$$

T is non-singular since (a_1,b_1) and (a_2,b_2) are linearly independent over $\mathbb{R}$. Thus if we let

$$\xi(x,y) = P \circ T^{-1}(x,y)$$

then

$$\xi(x,y) = xy(\alpha x + \beta y)$$

for some $\alpha, \beta \in \mathbb{R}$, and

$$\xi = J^3(\tilde{h} \circ T^{-1})(0) = J^3(\tilde{h})(0) \cdot J^3(T^{-1})(0) = P \cdot J^3(T^{-1})(0) .$$

Hence ξ is a 3-jet in the right orbit of P.

We observe that $\alpha\beta \neq 0$ since

$$\alpha = \begin{vmatrix} a_3 & b_3 \\ a_2 & b_2 \end{vmatrix} \neq 0 , \qquad \beta = \begin{vmatrix} a_3 & b_3 \\ a_1 & b_1 \end{vmatrix} \neq 0 .$$

Thus, the coordinate change $(x,y) \to (\frac{x}{\alpha}, \frac{y}{\beta})$, which is considered as a 3-jet in $L^3(2)$, moves the 3-jet $xy(\alpha x + \beta y)$ along its right orbit under $L^3(2)$ to the 3-jet $\xi_1(x,y) = \frac{1}{\alpha\beta} xy(x + y)$.

Then the linear map $(x,y) \to (\sqrt[3]{\alpha\beta}\, x, \sqrt[3]{\alpha\beta}\, y)$, considered as an element of $L^3(2)$, moves ξ, along its right orbit to the 3-jet $\xi_2(x,y) = xy(x + y)$. Hence P is equivalent in $J^3(2,1)$ to the 3-jet $\xi_2(x,y)$.

Now, the linear map $(x,y) \xrightarrow{\theta} (x + y, x - y)$ has matrix $\begin{pmatrix} 1 & 1 \\ 1 & -1 \end{pmatrix}$ (with respect to canonical basis) and it is an element of $L^3(2)$. Moreover, if we let $\xi_3 = \xi_2 \circ \theta$, then

$$\xi_3(x,y) = (x + y)(x - y)(2x) = 2(x^3 - xy^2) .$$

Finally, the map $(x,y) \to (\frac{x}{\sqrt[3]{2}}, \frac{y}{\sqrt[3]{2}})$ maps $\xi_3(x,y)$ to $\xi_4(x,y) = x^3 - xy^2 = g_6(x,y)$ in the list of Theorem 5.5 of Chapter 3. Hence P is equivalent in $J^3(2,1)$ to the 3-jet $x^3 - xy^2$.

We then observe that

$$x^3 = \frac{x}{3}(3x^2 - y^2) + \frac{y}{6}(-2xy)$$

$$x^2y = -\frac{x}{2}(-2xy)$$

$$xy^2 = -\frac{y}{2}(-2xy)$$

and

$$y^3 = -y(3x^2 - y^2) + \frac{3x}{2}(-2xy) ,$$

which imply that $m(2)^3 \subset m(2)\langle 3x^2 - y^2, -2xy\rangle + m(2)^4$, thus the 3-jet $x^3 - xy^2$ is (right) 3-determined. Hence $\tilde{h} \sim x^3 - xy^2$ and so f reduces (on the right) to $g_6(x,y) = x^3 - xy^2$.

Case (aII): $P(X,1)$ has one real and two complex conjugate roots. Then

$$P(x,y) = (a_1x + b_1y)(a_2x + b_2y)(\bar{a}_2x + \bar{b}_2y) ,$$

where $\bar{a}_2$, $\bar{b}_2$ are the complex conjugates of a_2 and b_2.

Consider the 2-form

$$\begin{aligned}(a_2x + b_2y)(\bar{a}_2x + \bar{b}_2y) &= |a_2|^2x^2 + (a_2\bar{b}_2 + \bar{a}_2b_2)xy + |b_2|^2y^2 \\ &= \alpha x^2 + \beta xy + \gamma y^2\end{aligned}$$

where $\alpha,\beta,\gamma \in \mathbb{R}$. Since (a_2,b_2), $(\bar{a}_2,\bar{b}_2)$ are linearly independent over $\mathbb{C}$, $a_2 \neq 0$ and $b_2 \neq 0$ so that α and γ are strictly positive. Hence, under a linear non-singular change of coordinates T, the quadratic form becomes $x^2 + y^2$. Hence the 2-jet $Q(x,y) = (a_2x + b_2y)(\bar{a}_2x + \bar{b}_2y)$ is in the same right orbit as the 2-jet $x^2 + y^2$, so we see that $P \sim (a_1x + b_1y)(x^2 + y^2)$.

Now, by an orthogonal change of coordinates, or by rotating coordinates, $a_1x + b_1y$ may be transformed into cx $(c \neq 0)$ and of course the form $x^2 + y^2$ does not change. So P is in the right orbit of the 3-jet $\xi(x,y) = cx(x^2 + y^2)$. Under the following linear change of coordinates

$$(x,y) \xrightarrow{\theta_1} \left(\frac{x}{\sqrt[3]{c}}, \frac{y}{\sqrt[3]{c}}\right)$$

$$(x,y) \xrightarrow{\theta_2} (\sqrt[3]{2}\,x, \sqrt[3]{2}\,\sqrt{3}\,y)$$

and

$$(x,y) \xrightarrow{\theta_3} \left(\frac{x+y}{2}, \frac{x-y}{2}\right),$$

we have

$$\xi \circ \theta_1 \circ \theta_2 \circ \theta_3(x,y) = x^3 + y^3 .$$

Hence, P reduces (on the right) in $J^3(2,1)$ to the 3-jet $x^3 + y^3$.

As in the Case (aI), $x^3 + y^3$ is (right) 3-determined, quite straightforwardly. Thus, f reduces to $x^3 + y^3 = g_5(x,y)$.

Cases (b) and (c) involve the same difficulty, namely, we can easily find a 3-jet to which P is equivalent, but such a jet is not 3-determined. In Case (b),

$$P(x,y) = (a_1x + b_1y)(a_2x + b_2y)^2 = \xi \circ T(x,y)$$

where $\xi(x,y) = x^2y$ and $T = \begin{pmatrix} a_2 & b_2 \\ a_1 & b_1 \end{pmatrix}$. In Case (c),

$$P(x,y) = (a_1x + b_1y)^3 = \xi \circ T(x,y)$$

where $\xi(x,y) = x^3$ and $T = \begin{pmatrix} a_1 & b_1 \\ 0 & 1 \end{pmatrix}$. In each case, the map ξ is not finitely determined (see Example 2.1 and Example 3.2 of Chapter 3). So we must include extra considerations for these two cases.

Case (b): Since f reduces to $\tilde{h}$, it is easy to see that $\operatorname{cod} \tilde{h} = \operatorname{cod} f \leq 4$, thus $\tilde{h}$ is finitely determined. In this case (Case (b)), we have

$$\pi_3(\tilde{h}) = P(x,y) = (a_1x + b_1y)(a_2x + b_2y)^2 \sim x^2y .$$

However x^2y is not finitely determined, so we must look at higher-order jets of $\tilde{h}$. By looking at $\tilde{h}$ and x^2y, one of which is finitely determined while the other is not, it follows that there must be a jet of $\tilde{h}$ which is not equivalent to x^2y. Let k be the largest number for which $j^k(\tilde{h}) \sim x^2y$. We shall show that $\tilde{h} \sim x^2y \pm y^{k+1}$. Without loss of generality, let $j^k(\tilde{h}) = x^2y$ and $j^{k+1}(\tilde{h}) = x^2y + \eta(x,y)$, where $\eta(x,y)$ is a homogeneous polynomial of degree $k + 1$, $k \geq 3$. Write out $\eta(x,y) = xy\varphi(x,y) + x^2\tau(x,y) + ay^{k+1}$ so that degree φ = degree $\tau = k - 1$. Consider a local origin preserving diffeomorphism Φ of $\mathbb{R}^2$ at $\overline{0}$ of the form $\Phi(x,y) = (x + \alpha(x,y), y + \beta(x,y))$ where α, β, to be chosen later, are homogeneous forms of degree $k - 1$. Then

$$\begin{aligned} j^{k+1}(\xi \circ P) &= x^2y + x^2\beta(x,y) + 2xy\alpha(x,y) + \eta(x,y) \\ &= x^2y + x^2\beta(x,y) + 2xy\alpha(x,y) + xy\varphi(x,y) + x^2\tau(x,y) + ay^{k-1} . \end{aligned}$$

Hence, if let $\alpha = -\frac{1}{2}\varphi$, $\beta = -\tau$ then

$$j^{k+1}(\xi \circ \Phi) = x^2y + ay^{k+1} .$$

Thus $j^{k+1}(\tilde{h}) \sim x^2y + ay^{k+1}$ for some non-zero real number a. One easily sees, from Theorem 3.1 of Chapter 3, that $x^2y + ay^{k+1}$ is $(k + 1)$-determined, so $\tilde{h} \sim x^2y + ay^{k+1}$. But clearly $x^2y + ay^{k+1} \sim x^2y \pm y^{k+1}$, where the sign depends on whether a is positive or negative. Now the codimension of either $x^2y + y^{k+1}$ or $x^2y - y^{k+1}$ is $k + 1$ (refer to Example 3.4 of Chapter 3). Since we require the codimension to be less than or equal to four, we must have $k = 3$. So, f reduces to $x^2y \pm y^4$, i.e. g_7 or g_{10} in the list.

<u>Case (c)</u>: In this case $j^3(\tilde{h}) \sim x^3$, which is not of finite codimension. Then $j^4(\tilde{h}) \sim x^3 + \eta(x,y)$, where η is homogeneous of degree 4 (perhaps $\eta \equiv 0$). We will show that this forces codim $\tilde{h} \geq 5$, which is beyond the agreed range.

For $\dim J^3(m(2)) = 9$, while $\dim j^{(3)}\langle\frac{\partial\tilde{h}}{\partial x}\rangle \leq 4$. This last inequality arises because $j^{(3)}\langle\frac{\partial\tilde{h}}{\partial x}\rangle$ is generated by $3x^2 + \frac{\partial\eta}{\partial x}$, $\frac{\partial\eta}{\partial y}$, x^3, x^2y. It follows that

$$\operatorname{codim} \tilde{h} = \dim_{\mathbb{R}} m(2)/\langle\frac{\partial\tilde{h}}{\partial x}\rangle \geq \dim_{\mathbb{R}} J^3(m(2))/\langle\frac{\partial\tilde{h}}{\partial x}\rangle \geq 5 .$$

Case (d): In this case $j^{(3)}(\tilde{h}) = 0$. Then $j^{(4)}(\tilde{h}) = \eta(x,y)$, where η is homogeneous of degree 4. A similar argument to that above shows that, in this case, $\operatorname{codim} \tilde{h} \geq 7$. Hence this case also cannot occur.

Thus we have obtained a complete list of germs such that any germ $f \in m(2)^2$ with codimension ≤ 4 will reduce to a germ in our list. Then, it is easy to check the codimensions and the coranks given in Table 5.2 and it is also clear that all the g_i are irreducible. Hence Theorem 5.5 of Chapter 3 is proved.

Having proved this basic classification theorem, we apply it to prove Thom's Classification Theorem. We shall prove Theorem 5.2 of Chapter 3; Theorem 5.1 is then a trivial corollary.

If F is a universal unfolding of f of codimension ≤ 4, it follows that $\operatorname{codim} f \leq 4$. Thus, we know, by Theorem 5.5 of Chapter 3, that f reduces (on the right) to a germ g which is one of the germs listed in Table 2 of Chapter 3, where either $g = g_0 \equiv 0$ or g is one of the 10 germs in Theorem 5.2 of Chapter 3. If $g = g_0$ then clearly F has a simple singularity at 0. If $g = g_i$, $1 \leq i \leq 10$, we need to calculate the universal unfoldings G_i of each of the germs g_i in this theorem. Having done so, we know that F will reduce to one of the canonical unfoldings G_i, $1 \leq i \leq 10$. We then need to insure that the index of reduction is zero.

How does one construct, for example, a universal unfolding G_5 of $g_5(x,y) = x^3 + y^3$? Recall that $\{x,y,xy\}$ forms a basis for $m(2)/\langle\frac{\partial g_5}{\partial \underline{x}}\rangle = m(2)/\langle 3x^2, 3y^2\rangle$. From the remarks in the opening paragraphs of this appendix, we know that

$$G_5(x,y,u,v,w) = x^3 + y^3 + ux + vy + wxy$$

is a universal unfolding of g_5.

The other universal unfoldings are constructed using the same method.

Now we must show that F reduces to the appropriate G_i with index 0. In fact, the reason that F reduces to G_i with index 0 follows exactly the line of reasoning given in Chapter 3 in explaining why reductions with index 0 are relevant in discussing unfoldings with local minima near 0.

To justify this remark, one needs to verify that each G_i, the universal unfolding of g_i, is an unfolding with local minima near 0. Let us look at G_5 first. Consider the unfolding of $g_5(x,y)$ (of codimension one)

$$H(x,y,t) = g_5(x,y) + t(x^2 + y^2).$$

Clearly H has local minima near 0. Since G_5 is a ($r\ell$-) universal unfolding of g_5, H is right-left induced from G_5. Hence there is $\lambda \in \varepsilon(1 + 1)$, $\Phi = (\varphi, \psi) \in \varepsilon(2 + 1, 2) \times \varepsilon(1,3)$ so that

$$H(x,y,t) = \lambda(G_5(\varphi(x,y,t),\psi(t),t)$$

and by the remarks in Chapter 3, $\varphi(x,y,0) = (x,y)$, $\lambda(s,0) = s$ for $s \in \mathbb{R}$, $(x,y) \in \mathbb{R}^2$. Thus for $\tilde{t}$ sufficiently near 0 and fixed,

$$H(x,y,\tilde{t}) = \lambda_{\tilde{t}} \circ G_5 \circ \Phi_{\tilde{t}}(x,y)$$

where $\lambda_{\tilde{t}}$ and $\Phi_{\tilde{t}}$ are diffeomorphisms of $\mathbb{R}^1$ and $\mathbb{R}^2$ respectively, and furthermore $\frac{\partial \lambda_{\tilde{t}}}{\partial s}(s) > 0$ for s near 0 in $\mathbb{R}$. Hence, for $\tilde{t}$ sufficiently small and fixed, $G_5(x,y,\psi(\tilde{t}))$ is right-left orientedly equivalent to $H(x,y,\tilde{t})$. Thus, $(\overline{0},\psi(\tilde{t}))$ will be a minimum for G_5 whenever $\tilde{t} > 0$ and sufficiently small. Note that this argument applies for any unfolding of $f \in m(p)^3$ where f is finitely determined.

Hence, each G_i is an unfolding with local minima near 0. Since F does so, it follows that F reduces orientedly to one of the G_i with index 0.

So, we have given in some, but not complete detail, most of the mathematical aspect of Thom's catastrophe theory. For further details we refer the reader to [82].

FURTHER READING

[1] V. I. Arnolʹd, Singularities of smooth mappings, Russian Math. Survey, Vol. 23, No. 1, Jan.-Feb. 1968, p. 1-43. (Translated from Uspehi Mat. Nauk 23, 1968, p. 3-44 .)

[2] V. I. Arnolʹd, Lectures on bifurcation in versal families, Russian Math. Survey, Vol. 27, No. 5, 1972, p. 54-123. (Translated from Uspehi Mat. Nauk 27, 5, 1972, p. 119-184.)

[3] V. I. Arnolʹd, Normal forms for functions near degenerate critical points, The Weyl Groups of A_k, D_k, E_k and Lagrangian singularities, Functional Anal. Appl. Vol. 6, 1972, p. 254-272. (Translated from Funkcional. Anal. i Priložen Vol. 6, No. 4, 1972, p. 3-25.)

[4] V. I. Arnolʹd, Classification of unimodal critical points of functions, Funkcional. Anal. i Prilozen. Vol. 7, No. 3, 1973, p. 75-76.

[5] N. A. Baas, On the models of Thom in biology and morphogenesis, lecture notes, Virginia (1972).

[6] J. M. Boardman, Singularities of differentiable maps, I.H.E.S. Math. 33 (1967) p. 21-57.

[7] J. Bochnak and T. C. Kuo, Different realizations of a non-sufficient jet, Indag. Math. 34 (1972) p. 24-31.

[8] J. Bochnak and S. Łojasiewicz, A converse of the Kuiper-Kuo Theorem, [61], p. 254-262.

[9] J. Bochnak and S. Łojasiewicz, Remarks on finitely determined analytic germs, [61], p. 263-270.

[10] T. Bröcker, Differentiable Germs and Catastrophes, Translated by L. Lander, London Math. Society Lectures Notes Series 17, Cambridge Univ. Press 1975.

[11] R. Courant and K. O. Friedricks, Supersonic Flow and Shock Waves, Interscience Publishers 1948.

[12] J. Damon, Topological stability in the nice dimensions ($n \leq p$), (to appear).

[13] E. Fermi, Thermodynamics, Dover Publications, 1936.

[14] D. H. Fowler, The Riemann-Hugoniot catastrophe and van der Waals equation, Toward a Theoretical Biology 4, p. 1-7.

[15] D. H. Fowler, Translation of [74] into English, Benjamin 1975.

[16] G. Glaeser, Fonctions composées differentiables, Ann. Math 77 (1963), p. 193-209.

[17] A. N. Godwin, Three dimensional pictures for Thom's parabolic umbilics, I.H.E.S. Math. 40 (1971), p. 117-138.

[18] A. N. Godwin, Topological bifurcation for the double cusp polynomial, Math. Proc. Camb. Phil. Soc. 1975, 77, p. 293-311.

[19] M. Golubitsky, An introduction to catastrophe theory and its applications, (to appear).

[20] M. Golubitsky and V. Guillemin, Stable Mappings and their Singularities, Springer-Verlag, Graduate texts in Math. 14 (1973).

[21] J. Guckenheimer, Bifurcation and catastrophe, Proc. Internat. Sympos. in Dynamical Systems (Salvador 1971) ed. M. Peixoto, Academic Press, N.Y. (1973).

[22] J. Guckenheimer, Catastrophes and partial differential equations, Ann. Inst. Fourier 23 (1973) p, 31-59.

[23] J. Guckenheimer, Review of [74], Bull. A.M.S. 79 (1973) p. 878-890.

[24] P. Hilton, Unfolding of Singularities, Colloquium on Functional Analysis, Campinas, Brazil, July 1974.

[25] K. Jänich, Caustics and catastrophes, Math. Ann. 209 (1974) p. 161-180.

[26] J. Kozak and C. Benham, Denaturation: An example of a catastrophe, Proc. Nat. Acad. Sci. 71, 1974, p. 1977-1981.

[27] N. H. Kuiper, C^1-equivalence of functions near isolated critical points, Sympo. Inf. Dim. Top., Ann. Math. Studies 69, Princeton Univ. Press 1972.

[28] T. C. Kuo, On C^0-sufficiency of jets of potential functions, Top. 8 (1969) p. 167-171.

[29] T. C. Kuo, A complete determination of C^0-sufficiency in $J^r(2,1)$, Inv. Math. 8, (1969), p. 226-235.

[30] T. C. Kuo, Characterizations of v-sufficiency of jets, Top. 11, (1972) p. 115-131.

[31] T. C. Kuo, The jet space $J^r(n,p)$, Proc. of Liverpool Sing. Sympo., p. 169-176. Lecture notes in Math. No. 192, Springer-Verlag 1971.

[32] T. C. Kuo, The ratio test for Whitney stratifications, [61], p. 141-149.

[33] T. C. Kuo and Y. C. Lu, On analytic function-germ of two complex variables, (to appear).

[34] H. I. Levine, Singularities of differentiable mappings, [61], p. 1-89.

[35] S. Łojasiewicz, Ensembles semi-analytiques, I.H.E.S. Math. 1965.

[36] E. Looijenga, Structural stability of families of C^∞-functions and the canonical stratification of $C^\infty(N)$, I.H.E.S. Math. Jan., 1974.

[37] Y. C. Lu, Sufficiency of jets in $J^r(2,1)$ via decomposition, Inv. Math. 10, 1970, p. 119-127.

[38] Y. C. Lu with S. H. Chang, On C^0-sufficiency of complex jets, Canada J. Math. Vol. XXV, No. 4, (1973) p. 874-880.

[39] B. Malgrange, The preparation theorem for differentiable functions, In "Differential Analysis" Bambay Colloq. 1964 Oxford, p. 203-208.

[40] B. Malgrange, Ideals of differentiable functions, Oxford Univ. Press, 1966.

[41] J. N. Mather, Stability of C^∞-mappings: I The division theorem, Ann. of Math. 87 (1968) p. 89-104.

[42] J. N. Mather, Stability of C^∞-mappings: II Infinitesmal stability implies stability, Ann. Math. 89 (1969) p. 254-291.

[43] J. N. Mather, Stability of C^∞-mappings: III Finitely determined map-germs, I.H.E.S. Math. 35 (1968) p. 127-156.

[44] J. N. Mather, Stability of C^∞-mappings: IV Classification of stable germs by R-algebra, I.H.E.S. Math. 37 (1969) p. 223-248.

[45] J. N. Mather, Stability of C^∞-mappings: V Transversality, Advances in Math. Vol. 4, No. 3, June 1970, p. 301-336.

[46] J. N. Mather, Stability of C^∞-mappings: VI The nice dimension, [61] p. 207-253.

[47] J. N. Mather, Notes on topological stability, Lecture notes, Harvard Univ., 1970.

[48] J. N. Mather, Right equivalence, Unpublished notes.

[49] J. N. Mather, Stratification and Mappings, Dynamical Systems, Academic Press, 1973.

[50] J. N. Mather, How to stratify mappings and jet space, (to appear).

[51] J. Milnor, Morse Theory, Ann. of Math., Studies 51, Princeton Univ. Press, Princeton, N.J. 1963.

[52] J. Milnor, Topology from Differential Viewpoint, The Univ. Press of Virginia, Charlettsville, 1965.

[53] J. Milnor, Singular points of Complex Surfaces, Ann. of Math. Studies 61, Princeton Univ. Press, Princeton, N.J., 1968.

[54] M. Morse, Relations between the critical points of a real function of n independent variables, Trans. A.M.S., 27 (1925) p. 345-396.

[55] M. Morse, The critical points of a function of n variables, Trans. A.M.S. 33 (1931), p. 72-91.

[56] Newsweek, Jan. 19, 1976, p. 54-55.

[57] L. Nirenberg, A proof of the Malgrange Preparation Theorem, [61], p. 97 - 105.

[58] V. A. Poénaru, On the geometry of differentiable manifolds, Studies in Modern Topology, Edited by P. J. Hilton, MAA Studies in Math. Vol. 5, p. 165-207.

[59] T. Poston and A. E. R. Woodcock, On Zeeman's catastrophe machine, Proc. Camb. Phil. Soc. 74 (1973) p. 217-226.

[60] T. Poston and A. E. R. Woodcock, A geometrical study of the elementary catastrophes, Lecture notes in Math. No. 373, Springer-Verlag, 1974.

[61] Proceedings of Liverpool Singularities Symposium I., Lecture notes in Math. No. 192, Springer-Verlag 1971.

[62] Proceedings of Liverpool Singularities Symposium II., Lecture notes in Math. No. 209, Springer-Verlag 1971.

[63] D. Ruelle and F. Takens, On the nature of turbulence, Commun. Math. Phys. 20, 1971, p. 167-192.

[64] D. Siersma, Singularities of C^∞-functions of right codimension smaller or equal than eight, Indag. Math. 25 (1973) p. 31-37.

[65] J. P. Speder, Equisingularite et conditions de Whitney, Nice I.M.S.P. 1972-1973.

[66] M. Spivak, Calculus on Manifolds, Benjamin, 1965.

[67] I. Stewart, The seven elementary catastrophes, The New Scientist, Nov. 20, 1975, p. 474-754.

[68] Symposium on structural stability, the theory of catastrophes and Applications in Sciences, Seattle 1975. Edited by P. J. Hilton, Lecture notes in Math. No. 525, Springer-Verlag 1976.

[69] F. Takens, A note on sufficiency of jets, Inv. Math. 13 (1971) p. 225-231.

[70] F. Takens, Singularities of functions and vectorfields, Nieuw Arch. Wisk. (3), XX (1972) p. 107-130.

[71] R. Thom, Ensemble et morphisms stratifiés, Bulletin AMS 75, 1969, p. 240-284.

[72] R. Thom, Les singularités des applications différentiables, Ann. Inst. Fourier 6 (1955-56), p. 43-87.

[73] R. Thom, Local properties of differentiable mappings, Differential Analysis, Oxford Press, London 1964, p. 191-202.

[74] R. Thom, Stabilité Structurelle et Morphogénèse, W. A. Benjamin, Inc., Reading Mass. 1972.

[75] R. Thom, Topological models in biology, Toward a Theoretical Biology 3, p. 89-116 and Top. 8 (1969), p. 313-336.

[76] R. Thom, A mathematical approach to morphogenesis: archetypal morphologies, Wistar Institute Symposium Monograph No. 9 (1969) p. 165-174.

[77] R. Thom, Modeles Mathématiques de la Morphogénèse, Ch. 1-3, mimeographed, I.H.E.S. (1970-71).

[78] G. N. Tyurina, Resolution of singularities of plane deformations of double rational points, Functional Anal. Appl. 4 (1970) p. 68-73. (Translated from Funkcional. Anal. i Priložen 4, 1 (1970) p. 77-83 .)

[79] A. N. Varčenko, Local topological properties of differentiable mappings, Izv. Akad. Nauk 38, 1974, No. 5 .

[80] R. J. Walker, Algebraic Curves, Dover, N.Y. 1950.

[81] C. T. C. Wall, Lectures on C^∞ stability and classification, [61], p. 178-206.

[82] G. Wasserman, Stability of Unfoldings, Lecture notes in Math. No. 393, Springer-Verlag 1974.

[83] G. Wasserman, (r,s)-stability of unfoldings, (to appear).

[84] H. Wergeland and D. ter Haar, Elements of Thermodynamics, Addison-Wesley Pub. Co., 1966.

[85] H. Whitney, The general type of singularity of a set of 2n - 1 smooth functions of n-variables, Duke Journal of Math., Ser. 2, 45 (1944) p. 220-293.

[86] H. Whitney, The singularities of smooth n-manifolds into (2n - 1)-space, Ann. of Math. 45 (1944), p. 247-293.

[87] H. Whitney, On singularities of mappings of Euclidean spaces I, Mappings of the plane into the plane, Ann. of Math. 62 (1955), p. 374-410.

[88] H. Whitney, Singularities of mappings in Euclidean spaces, In Symposium Internacional de topologia algebraica, p. 285-301, Universidad Nacional Autonoma de Mexico and UNESCO, Mexico City, 1958.

[89] H. Whitney, Elementary Structure of Real Algebraic Varieties, Ann. Math. 66 (1957), p. 545-556.

[90] H. Whitney, Tangents to an Analytic Variety, Ann. of Math. 81, 1965, p. 496-540.

[91] H. Whitney, Differentiable manifold, Ann. of Math. 37 (1936), p. 645-680.

[92] E. C. Zeeman, Geometry of catastrophe, Times Literary Supplement, 10 Dec. 1971.

[93] E. C. Zeeman, Differential Equations for the heartbeat and nerve impulses, Toward a Theoretical Biology 4, p. 8-67.

[94] E. C. Zeeman, A catastrophe machine, Toward a Theoretical Biology 4, p. 276-282.

[95] E. C. Zeeman, Catastrophe theory in brain modelling, Conference on Neural Networks, I.C.T.P., Trieste 1972.

[96] E. C. Zeeman with C. Isuard, Some models from catastrophe theory, Conference on Models in Social Sciences, Edinburgh 1972.

[97] E. C. Zeeman, On the unstable behavior of stock exchange, J. Math. Economics 1, 1974.

[98] E. C. Zeeman with D. J. A. Trotman, The classification of the elementary catastrophes of codimension ≤ 5, in [68], p. 263-327.

[99] E. C. Zeeman, Catastrophe theory: a reply to Thom, Manifold (Math. Inst. Univ. Warwick, 1974).

[100] E. C. Zeeman, Catastrophe theory, Scientific American, Vol. 234 No. 4, April 1976, p. 65-83.

[101] E. C. Zeeman, Euler Buckling, in [68].

[102] E. C. Zeeman, Breaking of waves, Symposium on Differential Equations and Dynamical Systems, Lecture Notes in Math. No. 206, Springer-Verlag.

INDEX

Graduate Texts in Mathematics

Vol. 14 Golubitsky/Guillemin: Stable Mappings and Their Singularities. 1973. xi–209. 10 illustrations.

Lecture Notes in Mathematics

Vol. 373 Woodcock/Poston: A Geometrical Study of the Elementary Catastrophes. 1974. v–275. 174 illustrations. paper

Vol. 393 Wassermann: Stability of Unfoldings. 1974. ix–164. paper

Vol. 525 Hilton (Ed.): Structural Stability, and Theory of Catastrophes, and Applications in the Sciences. 1976. vi–408. 90 illustrations. paper

Universitext • Hochschultext

O. Endler: Valuation Theory
H. Grauert/K. Fritzsche: Einführung in die Funktionentheorie mehrerer Veränderlicher
M. Gross/A. Lentin: Mathematische Linguistik
H. Hermes: Introduction to Mathematical Logic
H. Heyer: Mathematische Theorie statistischer Experimente
K. Hinderer: Grundbegriffe der Warscheinlichkeitstheorie
G. Kreisel/J.-L. Krivine: Modelltheorie—Ein Einführung in die mathematische Logik und Grundlagentheorie
H. Lüneburg: Einführung in die Algebra
S. Mac Lane: Kategorien—Begriffssprache und mathematische Theorie
J. T. Oden/J. N. Reddy: Variational Methods in Theoretical Mechanics
G. Owen: Spieltheorie
J. C. Oxtoby: Mass und Kategorie
G. Preuss: Allgemeine Topologie
B. v. Querenburg: Mengentheoretische Topologie
H. Werner: Praktische Mathematik I. (Ursprünglich erschienen als „Mathematica Scripta", Band 1)
H. Werner/R. Schaback: Praktische Mathematik II

Springer-Verlag New York Heidelberg Berlin